Writing creatively for work or study

Manchester University Press

Writing creatively for work or study

Easy techniques to engage your readers

Helen Kara

MANCHESTER UNIVERSITY PRESS

Copyright © Helen Kara 2025

The right of Helen Kara to be identified as the author of this work has been asserted in accordance with the Copyright, Designs and Patents Act 1988.

Published by Manchester University Press
Oxford Road, Manchester, M13 9PL

www.manchesteruniversitypress.co.uk

British Library Cataloguing-in-Publication Data
A catalogue record for this book is available from the British Library

ISBN 978 1 5261 7844 2 hardback
ISBN 978 1 5261 7845 9 paperback

First published 2025

The publisher has no responsibility for the persistence or accuracy of URLs for any external or third-party internet websites referred to in this book, and does not guarantee that any content on such websites is, or will remain, accurate or appropriate.

EU authorised representative for GPSR:
Easy Access System Europe, Mustamäe tee 50, 10621 Tallinn, Estonia
gpsr.requests@easproject.com

Typeset by Newgen Publishing UK

This book is for Lucy Pickering, a wonderful collaborator in work and play.

Contents

Figures vii
Tables viii
Creative elements viii
Acknowledgements ix

Introduction: creative writing in the workplace 1

1 Stories and fiction 20
2 Writing from life 44
3 Poetry 61
4 Graphic writing 85
5 Dramatic writing 112
6 Epistolary and digital writing 131
7 The personal is professional 149
8 Good practice in writing 165

Conclusion: drawing the threads together 183

On being autistic 187
Index 188

List of figures

Figure 0.1	Creative introduction to the author	15
Figure 3.1	Urban poetry (photo credit: Gary Butterfield/ Unsplash)	62
Figure 3.2	Shaped poem by Lewis Carroll	66
Figure 3.3	Shaped poem by Gaby Alvarado © Gaby Alvarado	67
Figure 3.4	Blackout poem	72
Figure 4.1	Cartoon by Marie Duval	87
Figure 4.2	Page 1 from *Conversation with a Purpose* by Helen Kara and Sophie Jackson	94
Figure 4.3	Page 2 from *Conversation with a Purpose* by Helen Kara and Sophie Jackson	95
Figure 4.4	Cover of issue 1 of the *Market Cafe Magazine* © Piero Zagami and Tiziana Alocci	97
Figure 4.5	*A Writer Writes* – comic by Helen Kara and Iqra Babar	101
Figure 4.6	Zine page 1 – cover page	102
Figure 4.7	Zine page 2	103
Figure 4.8	Zine page 3	104
Figure 4.9	Zine page 4	105
Figure 4.10	Zine page 5	106
Figure 6.1	Letter to a pre-scientist page 1	135
Figure 6.2	Letter to a pre-scientist page 2	136
Figure 6.3	Letter to a pre-scientist page 3	137

List of tables

Table 4.1   *Conversation with a Purpose* by Helen Kara   92
Table 8.1   Spreadsheet   170

List of creative elements

This is a writer   15
A story: 'The Competition'   36
Life writing: learning to read   57
Sending up a fibro flare   79
A comic   101
A zine   102
The start of a screenplay   126
A letter   144
A life-changing experience   159
I interview myself about writing   178

Acknowledgements

First and most importantly I am indebted to my beloved Nik for his unfailing support. I am thankful to Manchester University Press for taking on this book, and to Emma Brennan and Alun Richards and their colleagues for helping me through the process. I am also very thankful for my father, Mark Miller, who is always ready to talk about language, writing, and words – and, as a former professional indexer, kindly funded the index for this book. (He also tipped me off about Raymond Queneau. Thanks again, Poppa.) And I am grateful to the people who attend my creative professional writing seminars, workshops, and retreats because they have taught me so much over the years.

I also have some specific thank-yous to say:

Thanks to Jonathan Davidson for permission to reproduce his poem 'Metro' in Chapter 2.

Thanks to Mark Burstein, American expert on the works of Lewis Carroll, for helping some random woman from across the pond find a good enough image of 'The mouse's tale' for reproduction in Figure 3.2.

Thanks to Gaby Alvarado for permission to reproduce her shaped poem in Figure 3.3.

Thanks to Joan Larkin for permission to reproduce her sestina, 'Jewish food', in Chapter 3.

Thanks to Piero Zagami and Tiziana Alocci for permission to reproduce the front cover of the first issue of the *Market Cafe Magazine* in Figure 4.4.

Thanks to Leigh Forbes, voice artist and page-layout designer at Blot Publishing (https://blot.co.uk/), for invaluable help with the design of Figures 4.6–4.10.

Thanks to Jenelle Dozier from the US charity Letters to a Pre-Scientist (https://prescientist.org/) for permission to reproduce the annotated letter from Carly to Damien in Figures 6.1–6.3.

Introduction: creative writing in the workplace

Once upon a time ...

These words are a gateway to a fictional world, a story opening intended to stir excitement and anticipation. You may not have expected to find that well-known phrase at the start of this introduction. Perhaps seeing it here intrigued you, or maybe your heart sank, or you had some other reaction. Whatever your response, your emotions were engaged, and engaging readers' emotions is at the heart of writing creatively.

Have you ever read a business plan? An academic journal article? An evaluation or sales report? Did any of those documents engage your emotions? How memorable was the content? Of course there are exceptions, but for anyone who has read such documents, the answers to the last two questions might well be 'no' and 'not very'. I must have read dozens of annual reports in my life. I vaguely remember one which was a bit different: not A4 size, but a smaller, brightly coloured booklet. I can't remember which organisation produced it, or any of the content, but at least I can remember something about that document when all the others have blurred into fog in my memory.

Just like fiction, non-fiction writing communicates most effectively when it evokes emotion in its readers,[1] such as by demonstrating the writer's passion, creating tension, or using sensory language. The standard workplace document does not, generally speaking, affect its readers' emotions to any great degree. Yet research shows that activating readers' emotions makes content more engaging and memorable.[2]

Writing can be a powerful medium[3] for some individuals, institutions, and cultures. Within societies where writing is pervasive, the most powerful form of writing has become the dry and factual. This style is used for writing documents that are held to be important such as policies, laws, and contracts. Dry and factual writing is based on the idea that it is possible for language to describe or explain our world with precision. In fact, precision is difficult to achieve in any kind of writing, not least because the same word may mean different things to different people. Would you describe someone with a positive disposition as happy, cheerful, smiley, contented, or joyful? There are other options too.

In some languages one word can have different meanings. Take the English word 'fair'.[4] This has a range of meanings, including:

- treating people/animals/things equally;
- playing a game or a sport by the rules;
- light-coloured skin and/or hair;
- average or neutral (as in this scale: terrible, poor, fair, good, excellent);
- quite large (as in the statement, 'We've had a fair number of customers in today');
- pleasant, as an adjective for weather;
- just, in judgement;
- a pleasure ground with rides and attractions.

Then there is the homophone 'fare', which has three other meanings: the money you pay for a journey on public transport, food (as in, 'The restaurant serves Peruvian fare'), and how an individual or group gets on (as in, 'How did you fare in your job interview?' or 'Drivers fared badly over the holidays'). This can lead to ridiculous sentences such as, 'Although the weather was fair, we did not fare well at the fair because there was a fair number of people in each queue and the prices were not fair.' That said, although this sentence is grammatically correct, someone fluent in English would not construct that sentence unless they were writing a book and wanted an example. Instead they would choose synonyms to make the meaning clearer, such as, 'Although it was a nice day, we didn't enjoy the fair because the queues were long and the prices were high.'

As the American linguist Arika Okrent says, 'We have words that mean more than one thing, meanings that have more than one word for them, and some things we'd like to say that ... seem impossible to put into words.'[5] Another problem arises when professions adopt everyday words and give them specific meanings. Emergency medical dispatchers (EMDs) in the US used to have a script including a question to callers about whether the patient was alert. This caused confusion and delay, which is undesirable in a crisis response. It took some research to discover that 'alert' has a specific meaning in a clinical setting which is not shared by members of the public. Now the script has been rewritten to ask whether the patient is responding normally. This is much easier for most people to answer and still tells the EMD what they need to know.[6] (There is more about scripts in Chapter 5.)

This highlights a key point: we write for our readers or our audiences. Even a private diary is written for a reader: the writer's future self. Most other documents are intended for at least one other reader. The problem with the original script used by US EMDs was that it tried to communicate with ordinary people using professional jargon. Whatever you are writing – whether a script for a call taker, or a condolence card, or a screenplay – you need to keep your reader(s) or audience in mind and write for them. This may involve some imaginative work. If you are writing a conference presentation or an award acceptance speech, you may not know who will be in the audience; if you are writing an article or a comic for publication, you cannot be sure who your readers will be. In such cases you need to imagine who your audience or readers are likely to include and then write for them.

Some people think that writing to evoke a reader's emotions, and writing with precision, are mutually exclusive. Any competent poet can tell you this is an incorrect assumption. Even dull or difficult writing can evoke emotions in a reader such as boredom, frustration, or rage. In the business world, clear communication has generally been highly valued, though not always achieved. In academia, clarity has not always been valued, resulting in a kind of professional jargon written by academics for other academics. And in the non-profit sector, acronyms abound. At the time of writing there are over twelve hundred charity acronyms worldwide.[7]

Of course, there are professional contexts where technical or complex language is needed. However, the priority in most workplaces now is for more and better communication, driven by political agendas such as corporate social responsibility, public engagement, and widening participation. However, that doesn't mean that employees come ready-equipped with the skills to communicate their messages effectively to a range of audiences with different needs, such as policy-makers, practitioners, and the public.

Creative techniques can help any writer to make their work more accessible to more people. This book will show how techniques you might not think of in a workplace context, such as poetry or comics, can play a useful role in non-fiction writing. What is more, the book will explain how to approach and use these kinds of creative techniques. This is important because all writing requires creativity.[8] However, I am not suggesting that creative writing should replace standard non-fiction writing. My argument is that all writers would benefit from learning about these techniques, because they can make the process of writing more interesting and more fun. After all, if you don't enjoy writing, why do you think anyone would enjoy reading what you create?[9] These techniques can also enhance the final product to improve readers' experiences.

Some creative techniques are so subtle as to be almost imperceptible. A metaphor here, a contrast there – even these small touches can make your writing easier to read and understand. Other techniques are impossible to miss, such as a conference programme produced as a zine, or an executive summary written in iambic pentameter. The key to using creative techniques successfully in workplace writing is to know and understand these techniques and their effects on the reader. It also helps if you are clear about your own abilities and preferences. Some people would relish the prospect of creating a comic strip but balk at the idea of writing a sonnet; for others, the reverse would be true. Even so, in general, taking a creative approach to writing for work and study is something anyone can do.[10]

Creative writing in non-fiction

There are plenty of examples of dull non-fiction writing. Reach for any government policy document, instructions for an appliance, or a company's terms and conditions, and you are likely to find an example. It's harder to find examples of professional texts that make use of creative approaches, but they do exist.

Here's one example. Laura Stark studied the ways in which American institutional review boards (IRBs) make decisions. IRBs are committees which give or withhold ethical approval for proposed research projects as part of the business operations of academia. Stark collected archival data to use for context and observed a number of committee meetings to collect information about current decision-making practices. Sounds quite dry, doesn't it? Yet the last sentences of the introductory chapter in her book *Behind Closed Doors* read as follows:

> There is no dispute that the current research review system is flawed. On this everyone agrees: board members, administrators, and researchers. But the fact that IRBs provoke such heated debate is all the more reason to understand how these declarative bodies came into being and how they actually do their work. And so, on a dreary May afternoon, I stepped out of an elevator at Adams University Medical School, walked down a windowless corridor, and opened the conference room door where the IRB meeting was about to begin.[11]

In the final sentence, Stark uses a creative technique for two key purposes. First, she is managing a transition[12] from her brief introductory chapter, where she gives her readers an overview of IRBs – where they came from, what they are for, and how they work – to her first full-length chapter which is called 'IRBs in action'. In order to manage this transition, she uses a close-up[13] showing action in context, which gives her readers the sense of being with her on a journey. It is a brief but effective use of a creative technique at the micro level. (Other aspects of this excerpt are also notable, as we will see in the next chapter.)

At the macro level, Stark's book is written in a standard non-fiction style, which is not surprising as it is essentially an academic book. You can see a different approach in Saket Soni's book about human trafficking and the exploitation of migrants. The opening of *The Great Escape* is set in New Orleans and reads:

> When my parents called me the night of my twenty-ninth birthday, I was lying in wait for a human smuggler. I pried open my phone, barely feeling my fingers from the cold, and sent them to voicemail with a silent apology. I hadn't called home in months. This was regrettable, irresponsible. But they could sit tight for a few more hours. In Delhi, where they were, my birthday had already passed.

I was in an unheated car parked two blocks from the Home Depot on Claiborne, on an unlit street I couldn't name.[14]

This could be the opening of a crime novel or a psychological thriller. In contrast to Stark, Soni uses creative techniques throughout his book at micro and macro levels. A micro example is his use of sensory language ('lying in wait', 'barely feeling my fingers from the cold'). A macro example is that Soni sets some chapters of his book in various places in India, to introduce Indian workers before they migrate to the US and give an international context to these migrant workers' problems. Soni was born in India and writes eloquently about Indian society in places such as Kerala, New Delhi, Chennai, and Bihar, as well as writing about people's experiences in places like New Orleans, Pascagoula, Montgomery, and Greensboro in the US.

A macro example of creativity in business writing comes from iTunes, the Apple media library and player. The original iTunes terms and conditions were over twenty thousand words long, and they included paragraphs like this:

> The Service may also include certain features, functionality, and/or content that may be hosted, administered, supplied by or operated by third party service providers, such as social, community and public discussion areas, photo and video galleries, blogs, auctions, shopping, payment processing. These service providers may require that you agree to their additional terms, conditions, contracts, agreements and/or rules. Your compliance with any such additional terms, conditions, contracts, agreements and/or rules is solely your responsibility and will have no effect on your continuing obligation to comply with these terms and conditions.

Twenty thousand words is around one-third of the length of this book. I wonder how many people read all of those twenty-thousand-plus words before clicking 'I agree'. I bet it wasn't many. Later, Apple produced a shorter version of around seven thousand words, which is still quite long; a little longer than this chapter. Later still, in 2017, the original iTunes terms and conditions were published in graphic novel form, drawn by experienced graphic novel creator Robert Sikoryak. Sikoryak reproduced the full twenty-thousand-plus words, using Apple's founder Steve Jobs as his narrator. He also drew on the work of other renowned graphic novelists such as

Introduction

Alison Bechdel, and well-known cartoon characters such as Hulk and Snoopy, to make the unreadable entertaining. And it works.

It is also true that some fiction writers have used non-fiction techniques to good effect. For example, in his popular Discworld series of satirical novels, Terry Pratchett had a lot of fun with footnotes. In fact, the more we delve into this topic, the more ridiculous it seems to talk about writing techniques as though they belong to either fiction or non-fiction. It would make more sense to regard all writing techniques as available for everyone to master and use in ways that help to increase the effectiveness of written communication.

How creative writing techniques can help

So how can creative writing techniques help to make communication more effective? Essentially, by engaging readers' emotions, maintaining their interest, and activating their memories.

Readers' emotions may be engaged through devices such as identification, sensory language, surprise, or humour. Anyone who has felt they didn't live up to their parents' expectations at times – most people, probably – will identify with Soni's 'silent apology' as he sends his parents' call to voicemail. And sensory language need not be about pain: Stark's description of the May afternoon as 'dreary' and the corridor as 'windowless' may seem matter-of-fact, but she is also using the visual sense to create an atmosphere, leading us to suspect that the meeting she is going to will be monotonous, perhaps even oppressive. Terms and conditions produced in graphic novel form, or footnotes in a novel – even 'Once upon a time ... ' as an opening to a non-fiction book – evoke surprise because they are unusual. Then there is humour. Soni's opening sentence could be the start of a comic novel, so he swiftly moves to make sure the reader realises his text is serious by explaining that he is cold, not entirely sure where he is, and worried.

A great way to maintain readers' interest is to create and sustain tension. Soni has mastered this part of the craft, which is one reason his book is a compelling read. His opening creates great tension, because we want to know – of *course* we want to know – about the reality of human trafficking. Soni sustains this tension by taking his time in giving us the details. Delaying the reader's gratification is one way of building tension. Another is to introduce conflict. Soni includes interpersonal, international, and organisational conflict, though conflict could also be between competing ideas or arguments. Combining conflict and delay ramps up the tension. To do this, introduce the conflict, then work to maintain or ideally develop that conflict in your writing; don't resolve it until you have run out of other options. Tension is a key element in good storytelling and – here is an example in real time – we will get to that later.

If readers' emotions are engaged, and their interest is maintained, they are more likely to remember what they read. As a writer, do you want your readers to be intrigued by, interested in, and remember your work? If not, what is the point of writing for others to read? If you do, the techniques in this book will help. This introduction spells out some of *what* creative writing techniques can do for writers in workplace contexts and their readers; the following chapters cover *how* to create those effects.

What about truth?

There is a long-running and, so far, inconclusive debate about what constitutes truth and how to represent that truth in writing. A common misconception is that standard non-fiction writing conveys facts which equate to reality. Of course there is some truth in this, but it is not the whole truth. In fact it is impossible to use words to describe the whole truth of anything.[15] I can write 'a cup is on my desk' and that is true, there is a cup on my desk; even though you can't see it, you probably believe me, because a cup on a desk is quite an ordinary sight. But to tell the 'whole truth' of that cup would involve describing it in infinite detail: its colours and shapes, its history, how it was made, by whom, from what, where those materials came from, the number of times it has been used, when, for what, and so on. Writing can only communicate partial truth, and writers make choices all the time about what to include and what to leave out.

Having said that, some writing is more factual than other writing, and factual writing can be dry and tedious to read. If we leaven the facts with yeasty metaphors and serve them up with delicious descriptions, are we losing truth in the process?

Some thinkers have distinguished between two different kinds of truth. There are a number of terms for these, such as literal and real truths,[16] or fidelity and authenticity.[17] 'Literal truth' or 'fidelity' is the more representative, fact-laden kind of truth: there is a cup on my desk. 'Real truth' or 'authenticity' is the kind of truth we understand with our feelings as well as our minds. It is the truth in a good novel or a film that we say 'rings true', like a well-made bell, resonant and recognisable even though that novel or film may be pure fiction.

Creative techniques can help to add authenticity to non-fiction writing. Perhaps the May afternoon on which Stark went to observe her first committee meeting wasn't dreary at all. Maybe the sun was shining brightly. If so, her use of the word 'dreary' would lack fidelity; in one sense it would be untrue. Yet many readers could identify with the feeling of dreariness when walking down a windowless corridor to spend hours observing a committee meeting, whatever the weather might be doing outside. So Stark's use of the word brings authenticity to her account, engages the reader's emotions through identification, and thus makes her writing more effective.

These techniques are for arty people, though, right?

Wrong. In fact, quite insulting to scientists, accountants, engineers, and other such people, who are often capable of, and enjoy, writing creatively. For example, I know of a medical research laboratory in Perth, Australia, where the lab book is written in limericks.[18] Chemistry teachers in Portugal have been using poetry in the classroom to diversify learning processes and strengthen student development.[19] One of the earliest examples I have found of creative writing techniques in non-fiction is the use of dialogue to illustrate points in a book for novice engineers from 1973.[20]

Piper Harron studied mathematics at doctoral level at Princeton University in the US. Her dissertation abstract reads as follows:

> A fascinating tale of mayhem, mystery, and mathematics. Attached to each degree n number field is a rank $n-1$ lattice called its shape. This thesis shows that the shapes of S_n-number fields (of degree n = 3, 4, or 5) become equidistributed as the absolute discriminant of the number field goes to infinity. The result for n = 3 is due to David Terr. Here,

we provide a unified proof for n = 3, 4, and 5 based on the parametrizations of low rank rings due to Bhargava and Delone-Faddeev. We do not assume any of those words make any kind of sense, though we do make certain assumptions about how much time the reader has on her hands and what kind of sense of humor she has.[21]

The abstract of an academic dissertation or thesis provides a summary, and Harron's also reflects her creative approach. This is continued in the body of her dissertation where she writes separate sections for laypeople, her academic peers, and professors. Her dissertation contains poetry, dialogue, comic strips, diagrams, analogies, autobiographical details, and jokes. And equations: lots and lots of really complicated equations.

Harron succeeded in writing creatively even though she was writing about pure mathematics. Much of her writing is engaging and understandable even for non-mathematicians (the equations are more challenging, though they look pretty).

'Arty' people, too, can and do write creatively in workplaces for all sorts of different purposes. For example, there is a design and communications firm in Chicago, US, called Segura. They have produced a work-for-hire contract that reads:

TERMS
You give me money, I'll give you creative.
I'll start when the check clears.
Time is money. More time is more money.
I'll listen to you. You listen to me.
You tell me what you want, I'll tell you what you need.
You want me to be on time, I want you to be on time.
What you use is yours, what you don't is mine.
I can't give you stuff I don't own.
I'll try not to be an ass, you should do the same.
If you want something that's been done before, use that.

PRO BONO
If you want your way, you have to pay.
If you don't pay, I have final say.
Let's create something great together.[22]

This contract is creative in a different way. It is almost poetry: the repetition could be a literary device, and there are several rhymes.

Yet it doesn't appear to have been written intentionally as a poem. It looks as if it has been written using as few words as possible to say what needs to be said (which is of course what poetry does). It seems to have been created, not only to be engaging and understandable, but also to enable work to get done in the world.

This is a key function of many written documents: they are not just containers for words, but tools for people to use.[23] Documents such as contracts, wills, organisational policies and procedures, legal judgements, and terms and conditions provide levers for action. Yet, as the above examples show, there is still scope for creativity in writing and publishing these documents.

This chapter has discussed why we might want to use creative techniques in workplace writing. The rest of the book goes into detail about how and when to use these techniques. You may want to read the whole book, or you may be interested in some parts more than others. We start by looking at stories and fiction, because storytelling underpins all creative writing. Then there are chapters on writing from life; poetry; graphic writing (cartoons, comics, graphic novels, and zines); dramatic writing (play scripts, screenplays, and comedy writing); epistolary writing; and digital writing. Each of these five chapters starts by looking at how the forms are being used in workplace writing, moves on to discuss the relevant techniques, and includes exercises to help you use those techniques to enhance your own writing. Then there is a chapter on personal writing in professional life, another on good practice in writing, and a brief concluding chapter. To help you decide where you want to focus, here is a short overview of each chapter.

Chapter 1: Stories and fiction

Prose is the most widely used form in workplace writing and the most common way of writing conventional, fact-based text. All prose writing is creative because the writer is producing sentences and paragraphs that did not exist before. Non-fiction prose writers can take their creativity further by employing some of the devices used by fiction prose writers.

Most non-fiction writers use the device of storytelling, sometimes even without realising, because storytelling is a ubiquitous human habit. This chapter explains story structure and relevant writing devices: dialogue, repetition, recapitulation, metaphor, conflict, contrast, suspense or tension, and sensory language. It also explains the writer's 'voice'.

Stories can achieve a great deal in workplace writing, at least in part because of their relative familiarity. Where other forms are used, such as poetry, play scripts, screenplays, or cartoons, stories are often used alongside as container, interpreter, or enhancement.

Chapter 2: Writing from life

Life writing includes a very wide range of different types of writing. Conventionally thought of as autobiography, biography, and memoir, life writing also includes court proceedings, social media posts, case notes, personal diaries, and many more. As well as recording the lives of people, writing can also be used to record the lives of organisations, political parties, communities, and so on.

Life writing is based on observation and, although the writer may not be aware of this, usually involves some kind of theory. This chapter looks in detail at memoir, autoethnography, reports, and case studies, with examples to highlight creative aspects of writing in these genres. Examples are given of how snippets and fragments of memoir can be used creatively to enhance other types of non-fiction writing. The chapter covers the role of theory in life writing. It includes a discussion of the rights and wrongs of life writing when memory is unreliable. In short, we all lie to ourselves and others, and we can never tell 'the whole story'.

Chapter 3: Poetry

For some people poetry is unfamiliar, even daunting. Others see poetry as the preserve of poets alone and so not for them. Yet most of us loved nursery rhymes as children, and as adults we may adore a lyric writer who changes the way we think, or be grateful for a verse in a greeting card that says what we cannot. Far from being arcane and incomprehensible, poetry is an everyday activity and product that is all around us and accessible to almost everyone. And it is in use in workplace writing.

This chapter discusses what a poem is and outlines some of the simpler poetic forms as well as 'found poems'. Examples are given of ways in which poems have been, and are, used in business and academia. It also considers the relationship between constraints and creativity.

Chapter 4: Graphic writing

The use of cartoons, comics, graphic novels, and zines in the workplace is on the increase. We have seen that in 2017, iTunes published their terms and conditions in graphic novel form, and this form has also been used in academia. For example, the education researcher Nick Sousanis was the first person to create his doctoral dissertation in the form of a graphic novel, *Unflattening*. He was awarded his doctorate in 2014 and his book was published by Harvard University Press in 2015. Zines are less common in the workplace but, again, their use is increasing.

This chapter outlines the history of these formats, explains how they can be created, and gives examples of their use in the world of work.

Chapter 5: Dramatic writing

Play scripts, screenplays, and comedic writing are not as widely used in the workplace as some other formats, yet they have some important applications. Play scripts and screenplays are primarily constructed from dialogue and physical action, and take place in 'real time'. These forms are fundamentally dramatic, which makes them very useful for portraying interactions and relationships, and for engaging the reader's or audience's emotions. And comedy can be useful in a wide range of contexts.

This chapter gives examples of play-script and screenplay writing in the workplace. It explains how to write them and when they might be useful. It also introduces the hows and whys of comedy writing.

Chapter 6: Epistolary and digital writing

There are a range of other forms in use in the workplace. These include letters, emails, social media updates and comments, and writing for the internet – essentially any form with at least one intended reader. This chapter covers all of these formats, provides guidance on being polite online, and discusses the risks of creativity in the workplace.

Chapter 7: The personal is professional

Even those who would never consider themselves to be creative writers, storytellers, or poets, can benefit from using writing creatively in a personal capacity. This chapter discusses the emotional component of writing and introduces three roles writing can play: teacher, therapist, and friend. It also covers the potential of journals and diaries in the workplace.

Chapter 8: Good practice in writing

To use creative techniques effectively, it helps to have basic writing skills and processes in place. This chapter outlines the key skills you need such as understanding the use of grammar and how to structure your writing. It highlights and debunks eight common myths about writing.

Good practice also includes getting your writing done. This chapter reviews the barriers to and enablers of productivity, and suggests a range of practical actions for anyone struggling with putting words on the page. It also explains the importance of feedback, and of knowing who your readers or viewers are and what their needs might be.

There is also advice for people who are writing in a language which is not their first, and the chapter contains a discussion of when to be a little less creative.

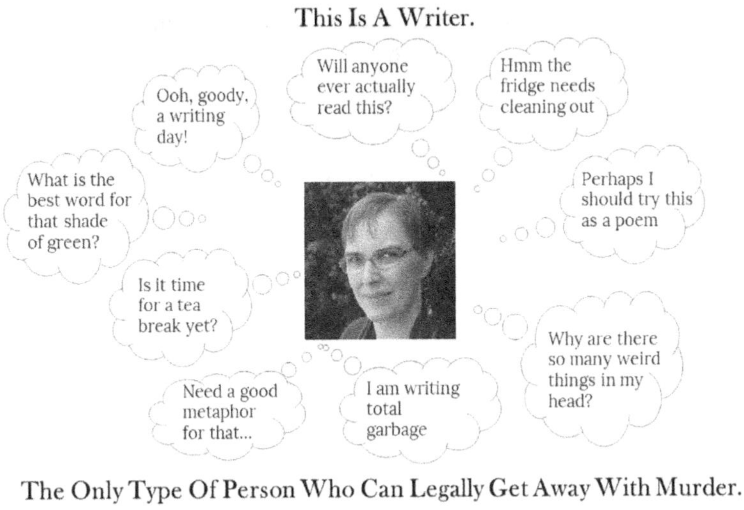

Introduction

This Is A Writer.

Thought bubbles surrounding a photo of the author:
- Ooh, goody, a writing day!
- Will anyone ever actually read this?
- Hmm the fridge needs cleaning out
- What is the best word for that shade of green?
- Perhaps I should try this as a poem
- Is it time for a tea break yet?
- Why are there so many weird things in my head?
- Need a good metaphor for that...
- I am writing total garbage

The Only Type Of Person Who Can Legally Get Away With Murder.

Figure 0.1 Creative introduction to the author

Try it yourself

This exercise is adapted from one in a book I wrote with Richard Phillips.[24] I cannot now remember which of us devised it; perhaps it was a joint effort. It has three parts.

Part 1: describe your workplace, in writing, as accurately and impartially as you can, as if you were an outsider. Use the present tense but do not use first person pronouns such as 'I' and 'we'. Do this before you read on.

Part 2: describe your workplace in writing again, but this time do use first person pronouns, to put yourself in the picture. Do this before you read on.

Part 3: compare your two descriptions. Is the first description really accurate and impartial? Do you think it is affected in any way by one or more of your own characteristics, for example, gender, race, ethnicity, social class? What does the second description reveal that the first conceals – and vice versa? Is there anything else worthy of note? Write down your answers, and consider what they tell you about how distance and closeness can affect your writing.

Notes

1 Sol Stein, *Solutions for Writers: Practical Craft Techniques for Fiction and Non-fiction* (Souvenir Press, 1995), p. 224.
2 Will Storr, *The Science of Storytelling* (William Collins, 2020), p. 45.
3 Peter Elbow, *Writing with Power: Techniques for Mastering the Writing Process* (2nd edn) (Oxford University Press, 1998), p. 279.
4 Thanks to Mark Miller for suggesting this example.
5 Arika Okrent, *In the Land of Invented Languages* (Spiegel & Grau, 2010), p. 11.
6 Becca Barrus, 'Precise language matters', *Journal of Emergency Dispatch*, 28 July 2021, www.iaedjournal.org/precise-language-matters [accessed 2 December 2022].
7 'Charity abbreviations', AllAcronyms, June 2021, www.allacronyms.com/charity/abbreviations [accessed 31 August 2024].
8 Elbow, *Writing with Power*, p. 11.
9 Patricia Leavy, *ReInvention: Methods of Social Fiction* (The Guilford Press, 2023), p. vii.
10 Elbow, *Writing with Power*, p. 9.
11 Laura Stark, *Behind Closed Doors* (University of Chicago Press, 2012), p. 8.
12 Scott Montgomery, *The Chicago Guide to Communicating Science* (2nd edn) (University of Chicago Press, 2017), p. 67.
13 Sam Leith, *Write to the Point* (Profile Books, 2017), p. 22.
14 Saket Soni, *The Great Escape: A True Story of Forced Labor and Immigrant Dreams in America* (Algonquin Books, 2024), p. 1.
15 Charles Lemert, *Social Theory: The Multicultural and Classic Readings* (Westview Press, 1999), p. 440.
16 Lucy Pickering and Helen Kara, 'Presenting and representing others: Towards an ethics of engagement', *International Journal of Social Research Methodology*, 20.3 (2017), pp. 299–309 (p. 299), https://doi.org/10.1080/13645579.2017.1287875
17 Michael Jackson, 'After the fact: The question of fidelity in ethnographic writing', in *Crumpled Paper Boat: Experiments in Ethnographic Writing*, ed. by Anand Pandian and Stuart McLean (Duke University Press, 2017), pp. 48–67 (p. 49).
18 Jillian Swaine, personal communication, 2017.
19 João Carlos Paiva, Carla Morais, and Luciano Moreira, 'Specialization, chemistry, and poetry: Challenging chemistry boundaries', *Journal of Chemical Education*, 90 (2013), pp. 1577–1579 (p. 1578), https://doi.org/10.1021/ed40030891

20 Dale Rudd, Gary Powers, and Jeffrey Siirola, *Process Synthesis* (Prentice-Hall, 1973), pp. 30–34.
21 Piper Harron, 'The equidistribution of lattice shapes of rings of integers of cubic, quartic, and quintic number fields: an artist's rendering' (PhD dissertation, Princeton University, 2016), abstract (unnumbered page).
22 Rusty Blazenhoff, 'One design firm's jargon-free contract: "Time is money. More time is more money"', BoingBoing, 22 June 2017, https://boingboing.net/2017/06/22/one-design-firms-jargon-free.html [accessed 28 March 2024].
23 Stuart Diamond, *Getting More* (Penguin, 2010), pp. 83–112.
24 Richard Phillips and Helen Kara, *Creative Writing for Social Research: A Practical Guide* (Policy Press, 2021), p. 30.

1
Stories and fiction

Introduction

Stories are essential to human life. The connection between people and stories is deep and powerful.[1] People all over the world share stories for education and entertainment, to create and develop relationships with one another, for collaboration and problem-solving. Although people have told stories since long before writing was invented, much of what we write today is in the form of stories, even if it doesn't look that way. A comic or an annual report, a documentary film script or an academic essay – whatever you are writing, it needs to tell a story. Being conscious of this fact, and working with it deliberately, will help you to communicate better through writing.

Whether oral, digital, or written, stories are created from words. That said, some stories are told using few or no words. Examples include the animated feature film *Belleville Rendezvous* (2003) written and directed by Sylvain Chomet from France, and the award-winning comic *Sky Rover* (2016) by Nunumi from Canada. Even though these works use few or no words in the telling, many words will have been written during their creation. And most stories comprise between dozens and hundreds of thousands of words.

Of course not everything that is written needs to tell a story. An organisational policy, the list of ingredients on a food or drink product, the 'help' pages of a software package – these kinds of documents exist to help us understand how things work and how to use them. We don't need stories about what happened when a company implemented their policy, or when someone ate or drank the product, or tried to use the software; we need clear and understandable information that we can apply for ourselves.

Most texts, though, benefit from some creative writing techniques, and many benefit from full-scale storytelling. The UK's Investor Relations Society offers an award for best annual report. The award description says: 'Best-in-class annual reports tell a holistic story'.[2] In this chapter we will review the creative writing techniques that can benefit non-fiction prose writing, and see examples of them being used in practice. First, let's think for a moment about truth.

Fiction is often identified with untruth. In the English language, 'story' is a synonym for 'lie'.[3] However, there are different kinds of truth. I discussed this in the previous chapter but it is an important point so worth considering again here. Two of the kinds of truth can be defined as literal truth, based on tangible evidence, and authentic truth, based on human experience.[4] I watched a film last night; that is literal truth, which could be confirmed by the person who watched it with me. The film, although entirely fictional, was gripping and moving, because its writing contained authentic truth. As we say, in a metaphor taken from bell-ringing, it 'rang true'. A cracked or distorted bell will make an unpleasant sound; for a bell to 'ring true' it must be whole and well made. Similarly, for our writing to convince, it must contain both literal and authentic truth within a well-made whole. Harvard's Professor of Literary Criticism, James Wood, calls this 'the reality of fictionality'.[5] This chapter will demonstrate how the reality of fictionality can be achieved through storytelling and the use of creative writing techniques.

Creativity in writing

All writing is creative. Even if you write about a dry topic packed with literal truth, you will select words and arrange them into sentences that nobody has ever written before. That is a creative process. Particularly so in a language such as English, which is rich in synonyms, so even the words you choose can affect the way your writing is perceived.

This is the start of an essay about the Kenyan writer and political analyst Nanjala Nyabola's fear of travelling alone to foreign countries in Africa.

> The pitch black night of the Sahara does not yield to the sunlight until it is good and ready, and when it does, it flees so fast that you would think the place is constantly bathed in blinding light. Stark sunrises turn the giant dunes dull brown for a scant few seconds; for a handful of minutes, as the sun is creeping up the sky, the sand glows. Then the sky cracks fully open and turns brilliant blue, and everything around you will shimmer in response.[6]

The author sets up the mood of the essay very effectively through clever and creative word choices. She uses verbs and adjectives like yield, flee, blinding, stark, creeping, cracks – words that combine to produce a subtle sense of danger. This conveys the authentic truth of her fear, layered over the literal truth that the sun rises daily over the Sahara.

If Nyabola had been writing about a road trip, she might have used different verbs, perhaps like this:

> The pitch black night of the Sahara does not give way to the sunlight until it is good and ready, and when it does, it accelerates so fast that you would think the place is constantly bathed in dazzling light. Sunrise turns the giant dunes dull brown for a scant few seconds; for a handful of minutes, as the sun is chugging up the sky, the sand glows. Then the sky opens fully and turns brilliant blue, and everything around you will vibrate in response.

As Nyabola wanted to communicate her emotional state, she chose subtly threatening words – and continued to build the spooky, ominous atmosphere in her next sentence: 'Until that moment when the blue scares off the dark, the dusty roads leading from Gorom-Gorom to Oursi, a small town outside a small town in northern Burkina Faso, are shrouded in the desert's secrecy, blanketed by inscrutable darkness and breathtaking silence.'[7] Here we have scares, shrouded, secrecy, inscrutable, darkness, breathtaking. As a reader you probably wouldn't make this kind of detailed analysis but, as a result of the author's word choices, you would be likely to pick up the feeling of dread. Perhaps this would generate a little excitement, too, because a terrified protagonist often makes for an interesting story.

So, we can be creative at the level of a single word. At the other end, where we reach the macro level, we can be creative about the structure of our work – unless this is prescribed for us, which it may be in some workplace contexts. For example, a manager or funder could require a report answering specific questions; most research reports still follow the IMRaD format (Introduction, Method, Results, and Discussion).[8] You may be able to think of similar examples from your own area of work or study. But even then you will be telling a story, albeit within a prescribed framework.

Story structure

One key point about stories is that they are mostly made from cause and effect. Young woman walks through forest to take food to sick grandmother, sees wolf in forest, visits grandmother, is eaten by wolf. That is not a story, it is a list of events – though you may recognise its kinship to an old European fairy tale. In that tale, *Little Red Riding Hood*, the young woman walks through the forest *because* that is the way to her grandmother's house. She meets a wolf on her way *because* wolves live in forests. She answers the wolf's questions about her errand politely *because* the wolf is a dangerous predator. The wolf races to get to her grandmother's house before her *because* he wants to eat her grandmother first and then her tasty self. Having eaten the grandmother, the wolf disguises himself as the grandmother *because* he wants to gain the girl's trust so she will come close enough for him to grab and eat her. And so on.

If you don't have a prescribed framework, then you can structure a story however you like. The simplest structure has three parts which may be known as beginning, middle, end, or context, action, results,[9] or set-up, confrontation, resolution.[10] The shortest example of a three-part story comes from the early twentieth century. Nobody knows who wrote this story. It has just six words:

For sale. Baby shoes. Never worn.

'For sale' sets the scene: somebody is selling something. 'Baby shoes' is what is being sold; what the story is about. 'Never worn' is the denouement which packs an emotional punch when we realise we have just read a six-word story about a baby who died.

This simple structure works with longer stories too, though in a longer story each section may need to be divided into sub-sections with differing purposes. There are many more complex structures too. Some people switch between personal stories and wider stories such as political or cultural stories. We will see examples of these in the next chapter. Others switch between present and past, such as the journalist Rebecca Skloot in her book *The Immortal Life of Henrietta Lacks*.[11] This is the story of a poor Black tobacco farmer in the US who died from cancer in 1951. Henrietta Lacks didn't know that doctors took some cells from her cervix, or that – unlike almost all human cells – they didn't die after a few days in a laboratory but proved to be immortal. Known as HeLa cells, they are now used worldwide to help with medical and other scientific advances, making millions for pharmaceutical and other corporations. For decades, Lacks' family had no knowledge of this and did not benefit in any way. This is a story of a family, life and death, exploitation, history, consent, and ownership. Skloot tells the story by moving backwards and forwards in time: the first chapter is set in 1951, the second in 1920–1942, then three more chapters in 1951, then the sixth is set in 1999, then back to 1951, and so on.

Then there are more unusual structures. Joan Wickersham uses an index structure for her book about her father's suicide and its impact on her family. The book is called *The Suicide Index*[12] and the contents page looks like an index until you realise the page numbers are sequential. In Padgett Powell's novel *The Interrogative Mood: A Novel?*[13] every sentence is a question. Some stories have epistolary structures, i.e., they are constructed from correspondence – letters, emails, text messages, and so on – which have great creative potential. Helene Hanff's *84 Charing Cross Road*[14] tells the story of a real-life friendship between an American booklover and a British bookseller through their edited correspondence. The story is so well told that it was adapted for the stage and for radio, and made into a film starring Anne Bancroft and Anthony Hopkins. (See Chapter 6 for more on epistolary structures.)

The structures of fiction can be a great asset to non-fiction writing. We met Saket Soni in the previous chapter. He is an Indian-born labour organiser, living in the US, who was instrumental in freeing Indian workers trafficked into a US labour camp. Soni wanted to write a book about this experience, and first he studied the techniques fiction writers use to create compelling, adrenaline-generating, page-turning stories.[15] Then he wrote his non-fiction book with the plot structure of a thriller and as much drama as he could infuse. *The Great Escape: A True Story of Forced Labor and Immigrant Dreams in America* was published by Algonquin Books in 2024.

I myself found writing academic journal articles much easier when I had become sufficiently skilled in writing fictional short stories to sell them to magazines. For your writing, you can choose to adopt or adapt an existing structure, or even invent your own. I have done that with this book, where each chapter ends with a creatively written example. (Given the number of books in the world, it is likely that others may also have used this structure, but I do not know of them.)

Writing devices

Storytelling is one device used by fiction writers which can also be helpful for non-fiction. There are many others too, such as: dialogue, repetition, recapitulation, metaphor, conflict, contrast, suspense or tension, and sensory language.

Dialogue

In 1973, Dale Rudd, Gary Powers, and Jeffrey Siirola published a book called *Process Synthesis* with the US publisher Prentice-Hall. This book was intended to teach novice engineers how to identify the best processing route to produce a product from specific raw materials. The work includes a range of skills, such as chemical reactions and thermal energy balancing. The authors declare their intent to 'present the material in a coherent and attractive form suitable for the students' first exposure to an engineering course'.[16] They do this, in part, by using dialogue. The dialogue is between 'Chemist' and 'Engineer' and it helps in two ways: by conveying information students need, and by modelling the differences between the two disciplines.

Dialogue generally does more than one job at a time. It can emphasise the fact that a story is being told, and can also serve to move that story forward. It is engaging for readers; I bet those engineering students appreciated the dialogue in *Process Synthesis*. Written dialogue can seem very realistic – even though it is usually highly contrived – because spoken dialogue is so common and understandable for most of us.

However, writing good dialogue is not as easy as having a good chat. If you listen to people talking to each other, you will hear a lot of 'filler' words and phrases such as 'um', 'er', 'so', 'like', and 'you know'. You will also hear half sentences, repetition, tangents ... in speech, these don't matter, because the purposes of real-life dialogue are different from the purposes of written dialogue. In real life, among other things, we converse to negotiate for what we want and to manage our relationships. In writing, dialogue exists solely to move the story forwards. Within that, it may also help in other ways, such as to develop character or plot, but if it is not also moving the story forwards, it shouldn't be there.

Imagine you decided to take coffee and cake to a friend and colleague in their office. When you got there, the dialogue might go something like this:

> 'Hi. I thought you could use a coffee.'
> 'Hey! How are you? Come on in, take a seat.'
> 'Thanks. I'm good, how are you?'
> 'I'm great. Is there cake in that bag?'

Contrast that rather dull and repetitive speech with the writing of American thriller author and master of dialogue Robert B. Parker. Spenser, a private detective and the story's hero narrator, is visited in his office by his friend and colleague Hawk. The short excerpt below is the start of a chapter.

> Hawk came into my office in the morning with some coffee and a bag of donuts.
>
> 'Coffee from Starbuck's,' he said. 'High-grown Kenya, bright and sweet with a hint of blackcurrant.'
> 'They sell donuts?'
> 'Naw, Starbuck's too ritzy for donuts,' Hawk said. 'Donuts are Dunkin'.'
> 'With a hint of deep fat,' I said.[17]

The first sentence, although it is not dialogue, acts as shorthand for all the usual 'hello, how are you?' parts of standard greetings.

So when we get to the actual dialogue we are straight into the action. Hawk's visit is unexpected, but he and Spenser take the time to share a little banter – which serves to assure the reader that their relationship remains strong – before Hawk gets to the point of his visit: Spenser is in danger because someone wants him dead.

The British fiction and non-fiction writer Nicola Morgan says this:

> The first thing to know about writing dialogue is that you should not try to write exactly as people speak ... At the same time, you can't write dialogue that the characters would never actually deliver. So, you devise a kind of stylised representation of speech, something that feels very natural.[18]

This is exactly what Robert B. Parker has done in the extract above: devised a kind of stylised representation of speech. He also follows the golden rule for written dialogue laid down by Sol Stein: use oblique responses, rather than the (mostly) straightforward responses of speech. Stein gives a useful example:

> Let's imagine a cocktail party at which a man is trying to come on to a woman he has just met. He might say:
>
> 'You are the most beautiful woman in the world.'
> Her instinct is to be polite. She might answer:
> 'Why, thank you.'
>
> This is boring. Nothing is happening. Watch what happens when her response is oblique:
>
> HE: You are the most beautiful woman in the world.
> SHE: I'd like you to meet my husband.[19]

Oblique dialogue moves the story forward more quickly and more dramatically. Robert B. Parker offers a delightful example soon after the extract above. Hawk has now told Spenser that 'somebody's looking to have you killed'.

> 'How much they paying,' I said.
> 'Now that's ego,' Hawk said.
> 'Well, how would I feel if somebody was offering five hundred bucks?'
> 'Be embarrassing, wouldn't it,' Hawk said.[20]

This short exchange carries a great deal of information. We learn from Spenser's first line that he is not particularly bothered to discover that his life is under threat, as he is more interested in the size of the contract. Hawk's oblique reply suggests that Spenser's identity may be fragile, and Spenser's own oblique reply is defensive, reinforcing this suggestion. Hawk's last reply is less oblique – he identifies the feeling Spenser is being defensive about – but it is not entirely straightforward, as it also serves to emphasise the point that, in their world, contract size is linked to status. By using dialogue so skilfully, Robert B. Parker can convey all of this in just twenty-nine words.

Another thing you may notice about Robert B. Parker's dialogue is that the only speech tag he uses is 'said'. Some authors make liberal use of other speech tags:

> 'No!' he screeched.
> 'Yes!' she snarled.
> 'I hope you die!' he bellowed.

Those speech tags are unnecessary because the dialogue alone makes it clear that the two characters are at loggerheads. Nicola Morgan points out that over-use of dialogue tags is lazy writing because it tells the reader what is going on rather than showing them.[21] Also, some speech tags are either ridiculous or tautologous. 'How funny!' she laughed – that's ridiculous (if this doesn't make sense to you, try laughing some words instead of speaking them). 'Hahahaha!' she laughed – that's tautologous. But 'said' is almost invisible to the reader.

Another option is to intersperse dialogue with actions or observations. This can be a much better way of showing the reader what is happening in your story. Here is an example, from the start of chapter 6 of *The Great Escape* by Saket Soni:

> Upstairs in my room shortly before midnight, still elated, I steadied my hand to pour Rajan a cup of tea. In return, he reached into a pocket. Out came a pouch made of a bright green handkerchief.
> 'I've been saving these for a special occasion,' he said.
> I untied it with care. It was full of eggshell-colored cashew nuts. As we munched and washed them down with malty Lipton, I shared the latest news.[22]

Stories and fiction 29

This ninety-eight-word passage contains only eight words of dialogue in a single speech. Yet it is a very good depiction of an interaction involving exchange and sharing – tea, cashew nuts, news.

Repetition

Repetition is a device to use with care. It is easy to overdo, whether by accident or on purpose, and that is always unhelpful. But when used judiciously, like strong seasoning, it can enhance the final product. Yet repetition can be misused at every level from word choice to overall structure. If you have ever complained that a book or a film is 'formulaic' or 'predictable', you were probably complaining about a structure that has been repeated too often. At the micro level, using the same word more than once in a sentence can lead that word to feel more like a speed bump than a word. Yes, I did repeat the word 'word' three times on purpose in that last sentence, to make the point! In between micro and macro, repetitive sentence structure is also problematic for the reader.[23] For example:

> The meeting took two hours and was attended by ten people. Everyone spoke at least once and some people spoke more frequently. The discussions were wide-ranging and the outcome was constructive.

This gets increasingly boring to read because each sentence has the same structure. The same information is more interesting (or, at least, less boring) with a more varied structure:

> The meeting lasted for two hours. It was attended by ten people, each of whom spoke at least once. Some spoke more frequently and the discussions were wide-ranging. The outcome was constructive.

So it is very important not to over-use repetition. Used carefully, though, repetition is an effective tool to create emphasis, or a mood, or a rhythm. Think of the song 'Happy Birthday to You', with its repetitive rhythmic lines which serve to emphasise the desire of all singers to celebrate and affirm the person having a birthday.

Synonyms are a great help in avoiding word repetition. Where there is no synonym, use repetition rather than linguistic contortions.

It is better to repeat an appropriate word than to use an inappropriate word. (This is one-third of the foundation of the BBC Radio 4 panel game *Just A Minute*[24] where panellists are challenged to talk on a topic for one minute without hesitation, deviation, or repetition. Apart from a few exceptions: it is acceptable to repeat the topic, and also short words such as 'I', 'and', or 'the'. This game can be very funny to listen to – or to try playing yourself – particularly when panellists are struggling to find a synonym for a word they have already used, aiming to avoid repetition without resorting to hesitation.)

Recapitulation

A sub-category of repetition is recapitulation, where a point made earlier in detail is summarised to help the reader understand the story. Comedians are masters of this art. A comedian may start their show by recounting a hilariously embarrassing experience, and mention in passing that they were wearing blue shoes. Then much later in the show, they can refer to a 'blue shoe incident' and the audience will laugh in recognition and understanding.

As a writer, you can use recapitulation in short or long pieces of work. In long pieces, it is useful to remind readers of points they may have read some time ago. In both long and short pieces it can be particularly useful in the ending. Effective story endings often echo the story's beginning in some way. This leads to a feeling of satisfaction and completion for the reader because the story has 'come full circle' (as the saying goes).

Metaphor

Metaphor is the representation of one thing by another. When I want to compliment or thank someone, I often say 'You are a star'. Of course, I am not suggesting that they are a spherical body of gas in outer space; I am using the word 'star' metaphorically to convey that I think what they have done is wonderful.

Metaphors are much more common in speech and writing than most people realise. Research has shown that we use a metaphor approximately every ten seconds in speech, and at a similar rate in writing.[25] Part of the reason we are unaware of the frequency is that some metaphors have lost their meaning due to over-use. These are known as 'dead metaphors' or 'conventional metaphors'. An example is 'front-line staff' used to describe hands-on workers in public services. This is so common we forget it is in fact a military metaphor, or indeed a metaphor at all; it no longer carries the connotations of going into noisy dangerous battle which would have made it a powerful metaphor at first. Dead metaphors are similar to clichés in that both are initially novel and impactful, so come to be used by many people, and therefore end up commonplace and banal.

On the other hand, 'creative metaphors', also known as 'unconventional metaphors', are newly or recently coined, and so have more power to make us think differently. The thought of inventing a new metaphor may seem daunting, but sensory language and imagery can help us find a way.[26] Let's suppose we want to find a new metaphor for loneliness. What does loneliness look like? Sound like? Taste like? Smell like? Feel like, to the touch? What animal would loneliness be? What temperature is loneliness? You may be able to think of other such questions which would be useful here, but we will start with these, and my answers:

> Loneliness looks like a solo mountaineer high above the snowline.
> Loneliness sounds like a sad bell ringing far away.
> Loneliness tastes like a dry biscuit in a dry mouth.
> Loneliness smells like a prison cell.
> Loneliness feels like rough stone.
> Loneliness would be an oyster, alone in its thick shell.
> Loneliness is cold.

From these answers we can build a metaphor for loneliness: she was as lonely as an oyster on a mountain top. Doubtless this is not the best metaphor for loneliness, but it is a good start. It is novel, visual for those who can visualise, and an oyster on a mountain top certainly would be lonely, far away from its usual environment and its fellow oysters in the sea. So the metaphor works. At least, it is my metaphor, so it would work in my writing. Your answers to these questions would, no doubt, be different, and so would lead to the building of a different metaphor.

Conflict

Conflict is essential for any compelling story because readers find it fascinating.[27] Fiction writers can create direct conflict between characters, though they may also use other types of conflict, such as conflict between races, religions, social classes, opposites, and so on. These types of conflict are also available to non-fiction writers. Conflict abounds in the workplace. There may be conflict between the priorities of different departments, organisations, sectors, or disciplines; between the wishes of customers/clients, staff, and shareholders or trustees; between the aims of operational and strategic staff, and so on. In the world of study there may be conflicting viewpoints, arguments, philosophies, et cetera. All these kinds of conflicts are potentially useful for writers.

Contrast

A relation of conflict is contrast,[28] which can also be used to good effect. If you are struggling to combine story and conflict, consider whether you can use contrast instead. Highlighting difference is interesting for readers. Thinking about opposites can help: informal/formal, known/uncertain, income/expenditure, and so on (though be careful not to let this lead you down the path to binary thinking, which is generally reductive and unhelpful).[29] To highlight differences effectively you need to be able to understand points of view which are different from your own.[30]

You can also use contrast within your writing, for example, by varying your sentence structure (as discussed above) or by using multiple voices. Fiction writers do this through the characters they invent, but non-fiction writers can do it too, in a variety of ways. One option is to use case studies, quotes, boxed examples, or other written illustrations in your stories. Another is to have sections written by different people, as in an annual report where the chairperson's report and the treasurer's report (or their equivalents) are usually presented in contrasting voices: the former often upbeat and optimistic, the latter clear and factual.

Suspense

Suspense, or tension, is another useful device which is particularly good at making writing compelling and keeping readers reading. If you have ever read a book that you found difficult to put down, which had you so engrossed that you missed your stop, bedtime, or appointment, you will understand the power of suspense. That was probably a fiction book, but suspense also has its uses in non-fiction writing. In the previous chapter we saw an example from *Behind Closed Doors* by Laura Stark which is also relevant here. The book recounts the author's research into the decision-making practices of IRBs, the bodies that give or withhold approval for proposed research in the US. Stark gathered documentary evidence and observed IRBs' meetings, then wrote a book about what she found. The book opens with a short introductory chapter, setting the scene for her research. As we saw in the introduction to this book, the first chapter of *Behind Closed Doors* finishes like this:

> There is no dispute that the current research review system is flawed. On this everyone agrees: board members, administrators, and researchers. But the fact that IRBs provoke such heated debate is all the more reason to understand how these declarative bodies came into being and how they actually do their work. And so, on a dreary May afternoon, I stepped out of an elevator at Adams University Medical School, walked down a windowless corridor, and opened the conference room door where the IRB meeting was about to begin.[31]

This is clever writing. We know there is a 'heated debate' which suggests conflict. And there's more …

Sensory language

At the very end of her chapter, Stark switches to sensory language. You have probably walked down a windowless corridor yourself, so the mention of one may provoke a memory. The combination of 'dreary' and 'windowless' with the prospect of a procedural meeting suggests some apprehension and we can all relate to that. Then the end of the chapter is what fiction writers call a cliffhanger. For sure it's not as dramatic as a situation where a hero is clinging by her fingertips to crumbling rocks a hundred metres above an ocean where hungry sharks are circling, attracted by the blood dripping from the flesh wound in her thigh … But Stark's ending is still a cliffhanger, and very likely to induce the reader to turn the page and carry on reading. Where they will find that the author does not dive straight into a description of the meeting, but backtracks into important contextual information, thereby adding a further layer of suspense.

Sensory language activates not only people's emotions but also their memories, which deepens readers' engagement with our writing. As we read a story, most of us experience the narrative as a developing model inside our heads[32] (this may not apply to some neurodivergent people, such as aphantasic and anaurelic people). Using sensory language helps readers to build their mental models more quickly, effectively, and fully. Active voice helps, too. Stark doesn't tell us she feels apprehensive about the meeting she is going to observe; she shows us her reluctance through her clever use of sensory language.

Combining the techniques

Here is an example which combines some of these techniques. It is taken from a chapter on structuring presentations in the business how-to book *Life's a Pitch* by Stephen Bayley & Roger Mavity:

> Chopin is greatly admired for his sonatas. The sonata form follows a very definite structure. A sonata begins with *exposition* (the composer setting out his idea), follows with *development* (the idea is developed and explored) and ends with *recapitulation* (the idea is re-expressed).
>
> The British army has been one of the more successful organizations in the last five hundred years of history. How they communicate is worth studying. The army doctrine on this is brilliantly simple: Say what you're going to say. Say it. Say it again.
>
> Look carefully at the army's refreshingly simple approach to communication and then look at the structure of a sonata. They're strangely similar, aren't they?
>
> Both start with a setting out of the idea; both then deliver that idea; both then summarize that idea.[33]

This is very clever creative writing. The authors use storytelling, repetition, and recapitulation to great effect in just 130 words. They tell a story of Chopin and sonata form, then they tell a story of the British army (which is in fact the same story), then they develop and explore the story ('Look carefully ... '). In their final sentence they recapitulate, telling the same story a third time in a different way. Bayley and Mavity use prose to exemplify the point they are making, by telling and re-telling a story, developing and exploring that story, and then restating it briefly in their own words. Which is exactly what they are advising their readers to do.

Voice

There are many different ways to tell any story. This is neatly illustrated by the French author Raymond Queneau in his book *Exercises in Style*, first published in 1947.[34] This is a collection of ninety-nine retellings of the same story, each in a different prose or poetic style. The story itself is quite mundane, with a narrator recounting a couple of slightly unusual experiences on public transport in Paris one day. It is the different styles which make the book interesting – so much so that it has been translated into dozens of other languages and is still in print.

Queneau's work aside, the best way to tell your story is *your* way. This means finding your own writer's 'voice', i.e., your tone and style, as well as the voice of the piece. This is as important in non-fiction as it is in fiction.[35] One good route to finding your own voice is to read your work aloud before you call it done. That will help you to uncover all sorts of errors, glitches, and clunky sentences which you could easily miss if you only read over your work by eye. A good route to finding the voice of the piece is to consider its readers or audience, and write for them. The difference between a writer's 'voice', the voice of the piece they are currently working on, and the voices of other pieces they have written and/or will write, is subtle but real.

Conclusion

It still seems novel to suggest that professional writing is – or should be – based on stories and that using story writers' techniques can benefit such writing. This is because of the perception that 'stories' are fictional and for entertainment, while 'work' is a serious business. Yet workplaces are full of stories. The organisational storytelling researcher David Boje defines a workplace as 'A collective storytelling system in which the performance of stories is a key part of members' sense-making'.[36] So surely it makes sense to use stories and storytelling techniques purposefully, to help us achieve our professional aims.

A story: 'The Competition'

'I have an exciting announcement for you.' Miss Lewis smiled at her class. 'We are holding a story-writing competition. There is a prize for the winning story and it will be published in the school magazine.'

Rebecca's heart sped up. She resolved to write her best story ever.

'Miss!' Shaun bounced in his seat. 'What's the prize, Miss?'

'A book token.'

Shaun folded his arms and scowled.

'The deadline is the end of this month, so you have three weeks to write your stories.'

'Do we have to write them at home?'

'If you want to enter the competition, Shaun, yes you do. But it's not compulsory.'

He furrowed his brow.

'You don't have to', Miss Lewis explained.

Shaun whooshed out a breath of relief.

'Does anyone else have any questions about the story-writing competition? No? Good. Turn to page 43 in your maths books.'

Rebecca bent her head over her book but her mind was full of words and there was no room for numbers. Maybe she would write about a magical car, like in *Chitty Chitty Bang Bang*, which could float and fly as well as driving. She could make it go underwater, too, like the Yellow Submarine. With a group of children, like in *Swallows and Amazons*, who go on adventures in the magical car. But children can't drive ... though if it's a magical car, maybe it doesn't need a driver. Maybe it can talk!

'Rebecca!'

Rebecca's head shot up. 'Yes, Miss Lewis?'

'I asked you, if you have a remainder of 1, do you round it up or down?'

Phew. An easy question. 'Sorry, Miss. You round it down.'

'Very good.' Miss Lewis moved on, and Rebecca tried hard to concentrate, but her mind kept wandering to her story. How many children should be in the group? How old would they be? What names should she give them? She couldn't wait to start writing.

It seemed an age until the bell went for home time. Rebecca's excitement rose at the prospect of the hours ahead. But first she had to get home safely. She picked up her satchel and went to the pegs to fetch her coat, eager to get out and find an adult to walk behind.

As she took her coat from its peg, Rebecca felt a sharp pinch on the back of her arm.

'Ow!' She whirled round to see Shaun.

'Keep your voice down.' He kicked her shin for emphasis.

Tears pushed at her eyes but she blinked them back, clenching her teeth together.

'That's better', he said. 'Now put your coat on like a good girl.'

She put one arm in a sleeve, wishing Shaun would evaporate or dissolve or something. As she twisted to find the other sleeve, he pinched the thin skin over her ribs.

'Shaun Smith! What are you doing?' Miss Lewis's face was angry.

'Nothing Miss!'

Rebecca could feel the tears flooding her eyes and silently begged them not to fall.

'Shaun, come with me.' As Miss Lewis led him away, Shaun glared over his shoulder at Rebecca, and mouthed, 'I'll get you for this.'

Her heart sank, then rose again as she realised her walk home would be safer than usual. Which meant she could think about her story instead of keeping a lookout all the time. She ran out and there was Amina and her Mum just heading out of the gates. Even better luck: she could walk close to them all the way to the corner of her street.

By the time she got home, Rebecca had her story half planned and was longing to start writing.

'How was school today?' asked Dad.

Rebecca put down her satchel. 'There's a writing competition!'

'What kind of writing?'

'Story writing.'

'Your favourite!'

'And a prize, and the winner gets published in the school magazine.'

She leaned on Dad and he put his arm around her shoulders. 'Give it your best shot, sweetheart.'

She gave him a hug and then picked up her satchel and went to her bedroom, eager to start writing. It was cold so she pulled a blanket from her bed and wrapped it round herself before she began.

Rebecca spent most of her spare time working on her story. She called it 'Tremendous Travellers' and enjoyed developing her characters: four girls who go on adventures involving lots of rescues, problems solved, and assistance provided, usually to hapless adults. The car was the only magical car in the world and could drive by itself, make decisions for itself and help the girls, though she decided she wouldn't give it the ability to talk after all because there was already a lot of dialogue to manage.

By the time her story was finished, it filled most of an exercise book. Waiting for the results of the competition was hard. Rebecca found it difficult to concentrate in school, and harder than usual to escape Shaun and his cronies. But she managed it, until one day when they caught her in the playground and made a ring around her.

'Time for you to get what you deserve', said Shaun.

Rebecca looked frantically from side to side but couldn't see a teacher anywhere.

One of the boys pushed her and she staggered, only for another boy to catch her and shove her in another direction.

'Good job', said Shaun. 'Let's all laugh so the teachers know how much fun we're having.'

Rebecca felt as if she was in a nightmare. All she could hear was the boys' laughter, and all she could feel was pushes and pulls, pinches and thumps.

'Stop, please. Please stop', she begged, but the boys took no notice and didn't stop until the bell rang for the end of playtime. By that time she was crying, her clothes torn and her body bruised. Instead of going back into school she ran out of the gates and straight home. Dad was out at work and she didn't know what to do. She wondered whether she should go back to school but the thought made her feel sick, so she slumped into the sofa and stared at the wall until Dad came home.

'What's up?' He dropped his bag by the door and came over to Rebecca. 'You're home early. How did you get that rip in your shirt? That's not like you, you know we're short of money.'

'I – ' Her throat closed up and she couldn't say any more.

He looked at her more closely, then sat down beside her and held her hand. 'Take your time', he said. 'Tell me when you're ready.'

She gulped. 'These boys – '

'At school?'

Rebecca nodded. 'They were playing a game with me.' Her voice wobbled and her throat closed up again.

'Not fun for you?'

She shook her head. 'It hurt', she whispered, the tears beginning to fall again.

Dad put his arms around her while she cried. His familiar smell of soap and tea was comforting. As her tears subsided he said, 'I'd better come to school with you tomorrow and have a word with the teacher.'

'Please don't, Dad.' Rebecca pulled away from him and sat bolt upright. 'That's how this all began. The teacher saw Shaun picking on me and gave him a telling off. That's why they went for me today. If you come into school it will only make it worse.'

'I can't let this go, Rebecca', he said.

'Dad? Could I – please, could I go to a different school instead?'

He thought for a moment. 'The one on the other side of town?'

'Yes.'

'It's a much longer walk – '

'I don't mind. Honest, Dad.'

He gave her a long look. 'Have you been thinking about this for a while?'

She nodded mutely.

'Leave it with me', he said. 'I will still have to come and talk to the teacher in the morning. I'll see what I can do.'

Walking to school with Dad the next day was fun. The sun shone and Rebecca skipped along beside him. If only he could stay with her all day... but of course that couldn't happen, the others would think she was a baby.

He left her in the playground while he went in to talk to the teacher before class. He came out just as the bell rang and gave her a smile and a thumbs up, then headed quickly off to work while she went into school.

'I have an announcement to make on this fine Friday morning.' Miss Lewis had her serious face on, and Rebecca wondered what was coming. 'About the story competition.'

Rebecca barely stifled a gasp. What with all the trouble she had forgotten all about her story.

'We have a winner', Miss Lewis went on. 'A very worthy winner.'

That doesn't sound like me, Rebecca thought, but then Miss Lewis said her name.

'Me, Miss? Really?' She could hardly believe her ears.

'Really, Rebecca.' Miss Lewis smiled at her. 'Come up and collect your prize.'

Rebecca walked to the front of the class and took the envelope and her story from Miss Lewis. 'Thank you', she said.

'Let's have a round of applause for Rebecca.'

Rebecca walked back to her seat as everyone clapped, and even the sight of Shaun doing a slow hand clap didn't spoil her pleasure.

At the end of school Miss Lewis asked her to stay behind when the class was dismissed. As soon as everyone else had gone, Miss Lewis said, 'I'm going to walk you home today.'

'Thank you, Miss', Rebecca said.

Rebecca saw Shaun lying in wait for her round the corner past the grocer's shop, but when he realised she was with Miss Lewis, he slunk away. She felt safer with Miss Lewis but she was glad to get home, and even gladder when Dad came home from work.

'Dad! I've got news!'

'So have I!' He grinned down at her. 'You go first.'

'I won the story competition!' She pulled the envelope out of her bag and waved it at him. 'I got a five pound book token!' She thought of all the worlds she could escape into and returned his grin.

'Well done!' he said. 'That's wonderful news, you clever girl.'

'What's yours, Dad?'

'Two things, actually. First, I have arranged for you to change schools.'

Rebecca bounced up and down. 'When?'

'Right now.'

'I don't have to go back?'

'You start at your new school on Monday.'

Rebecca whooped with delight.

'But that's not all', he reminded her. 'You're not the only clever one in this family. I got promoted to supervisor today!'

She gasped. 'Dad, that's amazing! More money?'

'Yes. We won't be so broke. In fact', he rubbed his hands together, 'I think we should celebrate. Fish and chips?'

Rebecca flung her arms wide, trying to express all the joy she felt. 'Dad! I think I'm living in a story!'

Try it yourself

You have two options here: one for fiction, one for non-fiction.

Option 1: fiction. Write a short story set in your workplace. Make sure it includes at least two characters, some dialogue, sensory language, and one or more emotional components. It is up to you whether you keep it entirely realistic or include fantastical or paranormal elements. Aim for five hundred to a thousand words, though you can write more, or less, if you wish. Choose your words carefully, and use the techniques outlined in the chapter where you can.

Option 2: non-fiction. Think of an experience from your workplace which is worth recounting; perhaps a story you have already told to others. Then write it down: again, use characters, dialogue, sensory language, and emotion. Aim for five hundred to a thousand words, though you can write more, or less, if you wish. Choose your words carefully, and use the techniques outlined in the chapter where possible. Try to tell the story as well as you can.

Whichever option you chose, once you have your first draft, read it through and consider what you could do to make it better. Does the narrative flow smoothly? Is the story satisfying? If not, what could you do to make it more enjoyable for readers?

Notes

1 Steven James and Tom Morrisey, *The Art of the Tale: Engage Your Audience, Elevate Your Organization, and Share Your Message through Storytelling* (HarperCollins Leadership, 2022).
2 'Best annual report', IRSociety, https://irsocietyawards.org.uk/awards/category/annual-report [accessed 29 July 2023].
3 'Story', Merriam-Wester Thesaurus, www.merriam-webster.com/thesaurus/story [accessed 7 July 2023].
4 Lucy Pickering and Helen Kara, 'Presenting and representing others: Towards an ethics of engagement', *International Journal of Social Research Methodology*, 20.3 (2017), pp. 299–309, https://doi.org/10.1080/13645579.2017.1287875
5 James Wood, *How Fiction Works* (Vintage, 2019).
6 Nanjala Nyabola, *Travelling while Black: Essays Inspired by a Life on the Move* (Hurst, 2020), p. 31.

7 Ibid.
8 Joshua Schimel, *Writing Science: How to Write Papers That Get Cited and Proposals That Get Funded* (Oxford University Press, 2012).
9 Paul Smith, *Lead with a Story: A Guide to Crafting Business Narratives That Captivate, Convince, and Inspire* (HarperCollins Leadership, 2012).
10 Reedsy Editorial Team, 'Story structure: 7 types all writers should know. 3. Three act structure', Reedsyblog, 8 August 2022, https://blog.reedsy.com/guide/story-structure/#3__three_act_structure [accessed 23 July 2023].
11 Rebecca Skloot, *The Immortal Life of Henrietta Lacks* (Pan Books, 2011).
12 Joan Wickersham, *The Suicide Index: Putting My Father's Death in Order* (Harcourt, 2008).
13 Padgett Powell, *The Interrogative Mood: A Novel?* (Serpent's Tail, 2011).
14 Helene Hanff, *84 Charing Cross Road* (Virago Press, 2002).
15 Wilfred Chan, '"We ferried 500 men out": How an organizer foiled one of America's biggest human trafficking operations', *The Guardian*, 10 March 2023, www.theguardian.com/us-news/2023/mar/10/the-great-escape-saket-soni-mississippi-human-trafficking-india [accessed 23 July 2023].
16 Dale Rudd, Gary Powers, and Jeffrey Siirola, *Process Synthesis* (Prentice-Hall, 1973).
17 Robert B. Parker, *Small Vices* (No Exit Press, 2008), p. 85.
18 Nicola Morgan, *Write to Be Published* (Snowbooks, 2011), pp. 151–152.
19 Sol Stein, *Solutions for Writers: Practical Craft Techniques for Fiction and Non-fiction* (Souvenir Press, 1995).
20 Parker, *Small Vices*, p. 86.
21 Morgan, *Write to Be Published*, p. 153.
22 Saket Soni, *The Great Escape: A True Story of Forced Labor and Immigrant Dreams in America* (Algonquin Books, 2024), p. 137.
23 Morgan, *Write to Be Published*, pp. 166–167.
24 *Just a Minute*, BBC programmes, www.bbc.co.uk/programmes/b006s5dp [accessed 6 December 2024].
25 Will Storr, *The Science of Storytelling* (William Collins, 2020), p. 44.
26 Owen Bullock, 'Poetry and trauma: exercises for creating metaphors and using sensory detail', *New Writing*, 18.4 (2021), pp. 409–420, https://doi.10.1080/14790726.2021.1876094
27 Morgan, *Write to Be Published*, pp. 109–110.

28 Bill Greenwell, 'Conflict and contrast', in *A Creative Writing Handbook: Developing Dramatic Technique, Individual Style and Voice*, ed. by Derek Neale (Black, 2009), pp. 14–27.
29 Meg-John Barker and Alex Iantaffi, *Life Isn't Binary: On Being Both, Beyond, and In Between* (Jessica Kingsley Publishers, 2019), p. 16.
30 Greenwell, 'Conflict and contrast', p. 17.
31 Laura Stark, *Behind Closed Doors* (University of Chicago Press, 2012), p. 8.
32 Storr, *Science of Storytelling*, pp. 28–29.
33 Stephen Bayley and Roger Mavity, *Life's a Pitch… How to Be Businesslike with Your Emotional Life and Emotional with Your Business Life* (Bantam Press, 2007), pp. 30–31.
34 Raymond Queneau, *Exercises in Style*, trans. by Barbara Wright and Chris Clarke (New Directions, 2013).
35 Morgan, *Write to Be Published*, p. 128.
36 David Boje, 'The storytelling organisation: A study of story performance in an office-supply firm', *Administrative Science Quarterly*, 36.1 (1991), pp. 106–126.

2

Writing from life

Introduction

Life writing is conventionally thought of as autobiography, biography, and memoir. In fact it is much broader, including a wide range of types and genres of writing. These include case notes, social media posts, personal journals, poetry, court proceedings, ethnography, parliamentary records, and many more. Life writing is not only used to record the lives of people, but also the lives of organisations, objects, political parties, geological periods – again, the range is very wide.

Let's start with the conventional. Autobiography is from the Greek for self-life-write. Memoir is from the Latin for memory. Sounds straightforward, doesn't it? Not so. People writing about life make mistakes and tell lies, omit and prevaricate, embellish and fabricate. One reason for this is that life writing is grounded in storytelling and, as Mark Twain said, you should never let the truth get in the way of a good story. Another is that people writing from life are often writing about other people, and those people (or their descendants) are likely to react to what has been written. People's reactions to what has been written about them are notoriously unpredictable: they may be delighted, furious, or despairing.

Roland Bannister found this out the hard way. His doctoral research centred on an Australian army band. He had previous experience as a soldier-musician, and was accepted by the Kapooka Band of Wagga Wagga to play in their trombone section. This enabled him to participate in, and observe, the life of the band and its members. Bannister could not anonymise his participants, as researchers often do, because Kapooka was the only Australian army band of its kind. So he took great care to tell the truth while leaving out anything that could embarrass his participants, damage their professional relationships, or jeopardise their careers. He shared drafts of proposed publications with anyone who was mentioned by name and asked whether they had any concerns. Mostly they did not – until one of Bannister's articles appeared in the *Australian Defence Force Journal* which was read by the band members' colleagues and superiors. This caused consternation for two of his participants, even though they had had the opportunity to read and comment on the article before its publication.[1]

It is arguable that all writing is from life; how could we write from anything else? Even the most fictional of fictions such as science fiction, fantasy, animated films, musicals, opera, or ballet appeal to people who recognise the characters, no matter how outlandish, dreamed up for us by their writers. For the writer, the unpredictability of people is an asset: characters may contradict themselves, be altruistic at one point and selfish at another, or generous and stingy, and so on – like all of us. This is helpful for writing with characters because it provides scope for the writer's creativity to roam free.

In this chapter I will focus on some of the aspects of life writing which are most relevant to writing in the workplace. These are: observational writing, memoir, autoethnography, and the use of theory, which all life writers do, whether they realise it or not.

Observational writing

It is also arguable that all writing is observational. As we write, either we are describing a phenomenon we have observed, or we are working from current or past observations of others' behaviour, others' writing, our own experiences, reactions, thoughts, and so on.

The Sayisi Dene people were Indigenous inhabitants of Duck Lake, Manitoba, a remote community in the northern forested interior of Canada. In 1956 they were forcibly relocated to slums on the outskirts of Churchill, a coastal frontier town several hundred kilometres away to the north-east. Almost one-third of the Sayisi Dene died in the next seventeen years. Then they reasserted their independence and began to relocate to Tadoule Lake, in the interior, around a hundred kilometres west of Churchill. Their story is told in a book written by Ila Bussidor, a former chief of the Sayisi Dene, and Üstün Bilgen-Reinart, a journalist. Üstün Bilgen-Reinart describes her first visit to Tadoule Lake in 1985:

> The twin-engine Beaver landed on the frozen lake at noon. I stood beside the plane, blinking in the blinding glare, with tiny icicles on my eyelashes. I had never been so far north before. I was astonished at the beauty of the place. The cold air was perfumed with spruce and wood smoke. A white light rose from the vast, snow-covered lake. The spruce forest loomed on the far shore. A fishing boat lay upside down on the beach. An esker (a sand hill left behind from an ancient glacier) rose beyond the beach, and log cabins were scattered here and there, beside teepees used for smoking fish and meat. Laundry, frozen stiff, moved in the wind.[2]

I have never been to northern Canada, nor to anywhere so far north, yet these words put a vivid picture in my mind. This is what the best observational writing does. (It does other things, too; we will come to those.) It is not only prose that can be beautifully observational, but also poems, plays, screenplays, comics – all forms of writing. Here is a poetic example, coincidentally also travel writing, by the English poet Jonathan Davidson:

Metro
Weeks later, I still think about Kyiv. I assume
It is still there, still opaque with people,
The same cafés, the same monuments,
The blocks of flats for workers, elderly
Lorries bringing things from somewhere.
It must still take five minutes by escalator
To reach the platforms. The roar of trains
Approaching is the turbulence of centuries.[3]

This poem was written after Davidson's visit to Kyiv in 2017, and was partly inspired by the Arsenalna metro station, which is the deepest underground station in the world, over a hundred metres beneath the city. His words, too, put a picture in my mind, though this time one influenced by my own visits to eastern European cities full of people, cafes, monuments, workers' apartment blocks, and elderly lorries. Davidson also makes a brief but powerful reference to the troubled history of Ukraine. Ukraine is surrounded by several European and Asian countries, most of which have taken a turn as ruler of part or all of the country we now call Ukraine. Poland and Lithuania, Austria and Russia, among others, tussled over the land for hundreds of years until it became the Ukrainian Soviet Socialist Republic, one of the founding republics of the Soviet Union, in 1922. It was occupied by Axis armies (Nazi German, Italian, and Japanese) from 1941–1944, and only achieved full independence in 1991. When he visited in 2017, Davidson knew of the fighting on the front line in Donbas, and could see that Ukraine was not a solidly established nation, and Russia was a threat.[4]

I have never visited Ukraine or Russia, but of course I was aware of the Russian invasion of Ukraine in 2022. I first read 'Metro' in December 2022, towards the end of the first year of that invasion, which meant the poem packed more of a punch for me as a reader than it would have done a year before. This is an interesting point to consider: sometimes a reader can take from a piece of writing something a writer didn't put there on purpose. This may be because of the reader's own experiences, or – as with Davidson's poem and my reading – because of changes in the world since the piece was written.

This has implications for us as writers. No writer – not even a keenly observant life writer – can control their readers' responses. No matter how much we worry about word choice, sweat over sentence structure, and wrestle with paragraphs, we can only control what we write, not what our readers read. As I wrote this book, I did plenty of worrying, sweating, and wrestling, in the full knowledge that some readers would dip in and out, skim-read, cherry-pick – and that's fine. It is in the nature of readers. When I am reading, I do the same.

As the verbs I chose in the last paragraph suggest, writing is an embodied process. This means it is physical – we write from and with our bodies – and also, inevitably, emotional. The French philosopher Roland Barthes wrote about the way creative outputs can puncture our composure to reach our emotions.[5] Barthes was writing about photography, but his point applies equally to writing and other creative arts. We cannot read, or write, without involving our emotions.

In fact, writing is a very emotional process. When I teach creative professional writing, I often ask students, at the start, to give me one word describing how they feel right now about their writing.

Answers vary from daunted, terrified, and overwhelmed, to happy, excited, and joyful. I point to the list of words they have produced and tell them they will each go through all of those emotions and more in the course of their writing careers. There is an apocryphal saying that writing is easy: you just sit down at your keyboard and open a vein. This portrays writing as painful, which it certainly can be. It can also be delightful, and everything in between.

Writing also, almost always, involves research. In observational writing, a writer's source material is the world around and within us: what we experience with our senses, externally and internally, and our records and memories of those experiences. We can observe and write directly about the place or situation we are in, which is useful for practice and for record-keeping. More often we will write about somewhere we have been, or something we have experienced, at a later date. Then we need to search or re-search our records and our memories. The internet can be useful here, too. It holds images, written accounts, audio files, and/or videos covering most parts of the world and much of human experience, which we can use to fill gaps in our understanding or memories.

Memoir

An autobiography is the story of the writer's own life told chronologically from birth to the present time. (A biography is the story of someone else's life, usually also told chronologically.) A memoir covers part of the writer's life and is more creatively structured than an autobiography, perhaps around one or more themes, or about a particular time period or experience.

I am focusing on memoir here because it is the most versatile of these classical life-writing genres. Memoirs have been in existence since ancient times: Julius Caesar wrote two military memoirs. There are numerous other genres of memoir: misery memoir, celebrity memoir, and so on. One of these is the workplace memoir. An early example of the genre from the UK is the series of books by James Herriot about his life as a veterinary surgeon in the Yorkshire Dales in the 1940s. These books mixed fact and fiction to excellent effect, putting Mark Twain's advice into practice, and became very successful. More recently, workplace memoirs have become more fact-based and hard-hitting, mostly in the realms of middle-class professions such as medicine and law.

Memoir may be short – a blog post, an essay – or take up several volumes. The story may be told in chronological order or using another kind of framework such as a thematic structure. Or, threads or fragments of memoir may be woven into other kinds of text, which is perhaps the most widely applicable option for workplace writing. I am doing that with this book: the illustrative pieces of creative writing at the ends of the chapters are all, in effect, forms of memoir.

David Spiegelhalter is a renowned Professor at Cambridge University, England, an eminent statistician, and an excellent writer. He has written an unusually readable book on statistics called *The Art of Statistics: Learning from Data*. This book is aimed at anyone who wants a better understanding, whether they are a student learning statistics, a professional using data, or a member of the public who wants to be more informed about how the media use (and misuse) statistics. The last section of the introductory chapter is headed 'This book', and it begins:

> When I was a student in Britain in the 1970s, there were just three TV channels, computers were the size of a double wardrobe, and the closest thing we had to Wikipedia was on the imaginary handheld device in Douglas Adams' (remarkably prescient) *Hitchhiker's Guide to the Galaxy*. For self-improvement we therefore turned to Pelican books, and their iconic blue spines were a standard feature of every student bookshelf.
>
> Because I was studying statistics, my Pelican collection featured *Facts from Figures* by M.J. Moroney (1951) and *How to Lie with Statistics* by Darrell Huff (1954). These venerable publications sold in the hundreds of thousands, reflecting both the level of interest in statistics and the dismal lack of choice at that time. These classics have stood up remarkably well to the intervening sixty-five years, but the current era demands a different approach to teaching statistics.[6]

Spiegelhalter's book is also a Pelican book. He doesn't say so, because he doesn't need to; it's on the cover. This is clever writing. The 'just three TV channels' in the first paragraph foreshadows the 'dismal lack of choice' in the second. (Of course not everyone would agree with Spiegelhalter's apparent position that more choice is a good thing. But it is his book, and he is putting forward his views.)

He portrays Pelican books as ubiquitous and best-selling – not a bad move when you're writing one! And he uses the snippet of memoir to lead the reader, in a gentle and engaging way, towards the key points he wants to make in the section: that statistics matter in today's world, that he will be dealing with statistical concepts rather than their technicalities, and that data literacy is a useful skill to acquire.

I have not conducted a forensic analysis of Spiegelhalter's book, but I think he only uses memoir that one time. He does include a lot of other real-world examples of statistics use, so it fits in well. Probably most readers wouldn't notice that he had introduced a fragment of memoir, they would simply enjoy his delightfully accessible writing.

Not many people want to write a full-length memoir, and even fewer are able to do so. But we can all use our own lives and experiences in similar ways to Spiegelhalter: to introduce, illustrate, or amplify points we want to make in our writing.

Some books benefit from a higher proportion of memoir. The journalist Sathnam Sanghera's first book was a full-length childhood memoir called *The Boy with the Topknot* about growing up in Wolverhampton, England, with Sikh parents who had immigrated from the Punjab. His first standard non-fiction book was called *Empireland* and was about the impact of the British Empire on the lives of people in Britain today. Not surprisingly, he used a lot of memoir in *Empireland*. The very first line reads 'My inbox at work is a nightmare.'[7] Chapter 2 is called 'Imperialism and me'. Chapter 3 begins with an account of the reading he has done for the book – and so on. This may sound as if I'm about to criticise Sanghera for being too full of himself, but in fact I think he has done an admirable job. He weaves his own experiences expertly through stories of imperial history and its impact on the present. Sanghera is honest about the emotional impact of looking imperial brutality straight in the face, while being unafraid to criticise his own community. Without his background in memoir, I am not sure he would have been able to take either of these steps. (I am far from alone in admiring this book; it was longlisted for a prestigious non-fiction prize.)

Memoir is an interesting blend of fact and fiction. The British chef Nigel Slater wrote a memoir of his provincial childhood called *Toast*. Reflecting on the twentieth anniversary of its publication, he said, 'Like most memoirs, there were details missing, some pieces exaggerated for effect, other stories simply forgotten. A few facts and names were fudged, if only to protect those involved.'[8] Memoir writing is, therefore, also a blend of factual and fictional writing. This blend is common in many types of writing, even writing we think of as purely factual or purely fictional. Though of course the proportion of fact to fiction can vary a great deal. We will consider this in more detail later in the book.

Autoethnography

Autoethnography is the academic equivalent of memoir, devised by American professor Carolyn Ellis in the 1990s. Where 'autobiography' comes from the Greek for 'self-life-write', 'autoethnography' comes from the Greek for 'self-people-write'. Ellis described it as 'an approach to research and writing that seeks to describe and systematically analyse personal experience in order to understand cultural experience'.[9]

Some memoirs are quite autoethnographic, and some autoethnographies are more like memoirs. The former is fine, the latter – from an academic point of view – not so good. Academics generally see research as sitting in between theory and practice: drawing from both, and then contributing to practice for sure and perhaps to theory too. Autoethnography is no exception. Memoir writers do not need to take this approach, though in fact some do link their own experiences with the flows and currents of society and culture. Sathnam Sanghera does this in *The Boy with the Topknot*. He writes about his father's and sister's schizophrenia, mostly by recounting personal experiences and memories, but also by weaving in wider social and cultural themes such as racial and gender differences and inequalities. Sanghera accompanies his father to visit his psychiatrist, also Indian, who explains that his father is doing well because he is able to live in a supportive Indian family, and says that white families usually break up when a member has a long-term mental illness. Sanghera comments:

> I thought this was just Dr Patel being patriotic. But it turns out he is right. Several studies have demonstrated that people who have schizophrenia in developing countries such as India, and people from such

cultural backgrounds, have a better chance of improvement than those who live in, or are from, the industrialized world.[10]

Throughout his memoir, Sanghera uses information from books and research on schizophrenia to link his memories, experiences, and his own journalistic research with wider issues. This is optional in memoir but necessary in autoethnography, although now and again a completely self-focused memoir somehow manages to get published as autoethnography. Most autoethnographers understand the point of switching focus, such as from the personal to the social, cultural, political, or theoretical. These perspective switches in writing work like camera view changes in film. Imagine a film with no close-ups, tracking, or long shots, with everything in medium view and the camera always static. It would be dull to watch. There are in fact over eighty types of film shot,[11] and the choice of shot is a crucial part of film-making, allowing the film-maker to emphasise or play down characters' actions and reactions, and so tell their story in the way likely to have most impact on viewers. Similarly, a writer can choose between a close-up on an incident from their own life, a tracking shot of a specific narrative within the overall story, a long shot looking at culture or theory, or a wide range of other perspectives from flashback to timeslip.

The writing of some autoethnographers is surprisingly cinematic. Jeffrey H. Cohen exemplifies this in the very first paragraph of his autoethnographic book about spending a year in a rural village in Mexico, with his wife Maria, doing anthropological fieldwork:

> Maria and I are sitting on the stairs of the porch of our house in Santa Ana del Valle, Oaxaca, Mexico. It's a warm afternoon in the late summer, August 1992. We haven't been in the valley very long, but Santa Ana is our home for the next year. We're moving into a house that Don Mauro (our patron) built for Jerónimo, his son who left for the United States in the 1980s ... The home has been empty for years, and it is dusty, filled with the detritus of poor harvests, unused bricks, and discarded furniture. Don Mauro describes our moving into his house as an event that will make it 'happy' and fill it for a while, and for the year, it will serve its purpose and not simply sit as an empty reminder that Jerónimo is not likely coming back. It doesn't take long to realize that there are a lot of empty homes in Santa Ana, and many Santañeros have left for other parts of Mexico and for the United States to seek their fortune ...[12]

Cohen begins with a close-up: a couple sitting on their porch stairs in a summer afternoon. Then he pulls back to show us their impoverished home, and then takes a long shot to make the point that poverty leads to loss as people leave a way of life that feels unsustainable to look for betterment elsewhere.

As often happens in good writing, the introductory paragraph does more work than simply describing a scene. It also introduces many of the book's key themes: family, culture, poverty/wealth, work, hierarchy, mobility, loss. These introductions are made so gently that someone reading the book for the first time would be unlikely to notice anything other than the evocation of a complex atmosphere of warmth and loss, happiness and emptiness. I certainly didn't; it was only when I went back to the book, to look more analytically at the writing, that I realised how cleverly the author wrote his first paragraph.

As with memoir, autoethnography does not need to occupy a whole publication. It can be used in snippets and fragments within a different format and structure. However, because autoethnography is only a few decades old, it has not yet made its way into other literary forms in the way that memoir has done. It clearly has the potential to do so. Perhaps you will be the writer to make that leap.

Reports and case studies

There are so many other kinds of life writing that it would not be possible to cover them all in a whole book, let alone a single chapter. So I will focus on two kinds of life writing which are particularly prevalent in the workplace: reports and case studies.

Reports range from a short progress report sent by email to a long and elaborate end-of-project report. The aim of a report is to give enough detail of what has happened, since an agreed start time or a previous report, to provide reader(s) with the information they need. Reports often seem quite dull to people who are not involved in whatever is being reported. This is a key reason why news reports tend to focus on deaths and disasters: the shock value makes them more compelling. Remembering that in writing a report or a case study, we are actually writing from life, can help us to make our own work more interesting for others to read.

For some people, the very idea of writing a report is daunting, and the suggestion that they might also take a creative approach makes them run away screaming (or at least feel like doing so). But, honestly, what makes writing more enjoyable for readers also makes it more satisfying for writers, both during the process and when receiving better feedback than they would otherwise have done.

For those who are at least willing to consider the idea, their next question, naturally, is, 'How?' In general, people are much less clear about how a report can be written creatively than how an autoethnography or a memoir can be written creatively. Yet they are all forms of life writing, so the answer is the same: the creativity comes from the writer.

At this point some more people run away screaming. Those who are left – of which, as you are still reading, you are evidently one – are more interested now. You may still be asking, 'How?' and the answer is: write *your* report, in the way only you can write that report. After all, as I have already argued, the very act of writing is creative. You are putting words together to make sentences, and sentences together to make paragraphs, that nobody else has ever produced before.

A report doesn't need to be written in a special style, or with lots of jargon, or long and complicated sentences. If you have a report to write, that report needs to be written in *your* style. This is often where the little voices in your head get louder: the ones that tell you you're no good, you can't do it, you shouldn't bother trying. Don't listen to those voices – and if you find that difficult, there are some tips to help you in Chapter 8. In the meantime, remember that those voices are not writers, but you *are* a writer, and you can take a creative approach to writing a report. This is what artificial intelligence cannot offer – and will never be able to offer.[13]

A report is a type of story. As the previous chapter showed, we are all steeped in stories. You know how to tell a story – you have probably told and heard several in the last twenty-four hours alone. And you know what makes a good story, from all the films, novels, plays, comedy shows, and so on that you have experienced in your life. So you can use that knowledge to create a good report.

Case studies can be useful when you are writing reports, though they also have a role in many other kinds of documents such as textbooks, policy documents, academic works, and newspaper columns. A case study is a written representation of one person, or organisation, or incident – any single phenomenon or 'case'.

There are different types of case study, such as medical case studies which can be very long and detailed. There is also a research method called the case study, which involves making a thorough examination of a phenomenon from a range of perspectives and in a variety of ways.[14] Here, though, I am talking about illustrative case studies, usually used within a longer document such as a report, thesis, or book. These kinds of case studies should have a clear focus and be written using vivid language to make a picture for the reader. There are no rules about case study length apart from 'as short as possible' which may, in fact, be quite long.

While reports are generally required to be factual, case studies may need to be fictionalised, at least to some extent. Of course if you have permission to write a factual case study of an identifiable person or group, then that's great. If you don't have that kind of permission, there are other options. If you were writing a report of your work which involved visiting people's homes, you might take a composite approach to a case study. You could do this by incorporating various elements from different visits, to form a single case study which is authentic while protecting the anonymity of your clients.[15] Alternatively, if you were writing a report of your work as a hospital nurse, you might draw on your experiences to write an entirely fictional, but still authentic, case study.

The role of theory

Whether or not the writer is aware of this, writing never happens without theory – or theories. This book is based on the theory that a certain set of skills are required to write for professional reasons, regardless of whether that writing is for work or study, or is predominantly factual or fictional. Sathnam Sanghera's memoir *The Boy with the Topknot* was based on the theory that there is a relationship between mental ill-health, ethnicity, and culture. The theory of many romance novels could be summed up as 'love will find a way'.

Some people regard theory as complicated and abstract, probably because they have seen or heard some other people using theory in complicated and abstract ways. Indeed, theory can be conceptually complex, but in essence, a theory is simply a way of looking at the world. We all have our own theories, such as that certain categories of people are mostly well-intentioned, or scary, or interesting, or dishonest. And we can make use of other people's theories when they are helpful for our own work. Memoir like Sanghera's and autoethnography like Cohen's demonstrate the usefulness of theory in life writing – and there are many other examples too.[16]

The theory of a piece of writing is like its compass: it helps the writer to find the direction in which their writing needs to go. Not all writers are aware of theory, though they are likely to use theory nevertheless. One of my own theories is that writers write better if they use theory consciously and deliberately. This can be done in many ways and with varying levels of complexity. At its simplest, though, it means articulating the theory guiding their own work, and then using that theory to steer them as they write.

Something we don't have is a fully developed and clearly articulated theory of creative writing itself.[17] This may be because writing is widespread but not universal; individual and cultural; historical and current – too complex in itself to theorise easily. It may also be because learning to write well enough so that others will want to read our work is a complex process. There is no right way to do this, though there is a right way for *you*. You may do your best writing in the morning, or the afternoon, or the evening. Your ideal length of time for a writing burst is likely to be different from someone else's. You might be more motivated by a word goal or a time goal. You may prefer to start with the section that feels easiest or the part that feels hardest. Of course there are huge numbers of combinations of these variables, and they are not the only variables, which demonstrates what a complex process writing can be. The best way to find your own right way to write is simply by trying different options to discover which ones work for you.

Rights and wrongs in life writing

No writing is value-free, and life writing is perhaps more fraught with moral and ethical dilemmas than most other kinds of writing. Questions any life writer will have to answer include:[18]

1. How much of the truth should I tell?
2. How can I avoid my work causing harm?

3. How can I ensure my work maximises its potential to do good?
4. Am I writing within all the laws that apply to me and my work?
5. Am I going to upset people I know by including them in my story?

Life writers will also benefit from understanding 'reality' as a set of multiple, overlapping, conflicting perspectives, rather than a single verifiable thing. You have probably had the experience of telling a story from your life to a group of family or friends, and being interrupted by someone who says, 'That is not how it was', and then shares their memory which differs from yours. There is often no way to prove which version is right – and it may be neither, for several reasons. First, human memory is notoriously unreliable.[19] We lie, to ourselves as well as to others, for all sorts of reasons: shame, ignorance, wanting to protect other people's feelings, and so on. And, inevitably, we can't tell 'the whole story'.[20] As we have already seen, it is impossible to write down everything about anything, even a simple object like a cup. So we have to choose the stories we tell, and in choosing which story to reveal, we are also, inevitably, choosing which stories to conceal. We don't often think about this, we usually stay focused on the stories we are choosing to tell – but for ethical life writing, we also need to be aware of the stories we are not telling, and why we are not telling those stories.

Gaining this awareness requires some reflexive thinking. Reflexivity is a word with more than one meaning. Here I am using it in the descriptive sense where it means 'able to reflect', i.e., to take a metaphorical step back from your writing and think about what you are doing and why. This is an essential skill for ethical life writing. The act of life writing is a form of exploration of the past, and that exploration affects how we remember, think, and feel about the past. This means we need to check in with ourselves from time to time to make sure we are still on the right track.

Life writing: learning to read

I learned to read when I was three years old. My mother taught me; she had recently given birth to my baby sister, and I suspect she badly needed to find a way for me to occupy myself. This was 1967 and we did not have a television. I had some toys, but playing on my own was only fun for a little while. My mother loved to read – my father too; one of them would always read me a bedtime story, and we made regular visits to the local library.

My mother's teaching method was based on bribery. If I read a whole page – probably just a few words in the books I was learning from – she gave me a Smartie, a small brightly coloured chocolate sweet. When we reached the end of a whole book, we went to the corner shop and she bought me an orange ice lolly. At the age of three, I was highly motivated by Smarties and orange ice lollies, so this approach worked well for me, and I soon learned to read story books and enjoy them so much that the enjoyment became its own reward. No doubt that was a great relief to my mother – though I think finding a way for me to occupy myself was not the only reason she taught me to read.

Reading would of course help me at school and at work, but my mother knew she was giving me much more than a useful skill. She was giving me insights, learning, role models, laughs, and whole new worlds I could spend time in when my own world wasn't fun. She passed on to me – and, later, to my sister – her love of books. In later years, when I became a successful writer myself, my mother was very proud and happy. As was my father. So I dedicated my first full-length sole-authored book to 'my parents ... who taught me to read, write, and think, gave me a love of learning, and encouraged me even when my choices were different from their own.'

Try it yourself[21]

Think of an experience you would prefer to keep private: perhaps something embarrassing, or sexual, or incriminating. Then answer these five questions, in writing, in as much detail as you can:

1. Why would you not choose to write about this experience?
2. What might you learn if you did write about your experience?
3. If you wrote about it, and your writing was published, how could that affect you and other people?
4. Is there any possibility of positive outcomes from this? If so, what might they be?
5. What might other people learn from reading about your experience?

Writing from life

These are the kinds of questions that life writers often have to grapple with, even if they are writing a report or a case note. Working through them for yourself will give you more insight into the processes and pressures of life writing.

Notes

1. Roland Bannister, 'Beyond the ethics committee: Representing others in qualitative research', *Research Studies in Music Education*, 6 (1996), pp. 50–58.
2. Ila Bussidor and Üstün Bilgen-Reinart, *Night Spirits: The Story of the Relocation of the Sayisi Dene* (University of Manitoba Press, 2006), p. xiii.
3. Jonathan Davidson, 'Metro', in *A Commonplace: Apples, Bricks, and Other People's Poems*, ed. by Jonathan Davidson (Smith|Doorstop Books, 2020), p. 73.
4. Jonathan Davidson, personal communication.
5. Roland Barthes and Richard Howard, *Camera Lucida: Reflections on Photography* (Vintage Classics, 1993).
6. David Spiegelhalter, *The Art of Statistics: Learning from Data* (Penguin Random House, 2019), p. 16.
7. Sathnam Sanghera, *Empireland* (Penguin Random House, 2019), p. 1.
8. Nigel Slater, '"My little book sprouted legs": Nigel Slater on how his memoir Toast became a phenomenon', *The Guardian*, 22 January 2023, www.theguardian.com/food/2023/jan/22/nigel-slater-how-his-memoir-toast-became-a-phenomenon [accessed 29 July 2023].
9. Carolyn Ellis, Tony Adams, and Arthur Bochner, 'Autoethnography: An overview', *Forum: Qualitative Social Research*, 12.1 (2011), Art. 10, p. 1.
10. Sathnam Sanghera, *The Boy with the Topknot: A Memoir of Love, Secrets and Lies in Wolverhampton* (Penguin Books, 2008), p. 125.
11. 'Types of film shots: 80+ shots you must know', Nashville Film Institute, www.nfi.edu/types-of-film-shots/ [accessed 29 July 2023].
12. Jeffrey H. Cohen, *Eating Soup without a Spoon: Anthropological Theory and Method in the Real World* (University of Texas Press, 2015), p. 1.
13. David Brooks, 'In the age of A.I., major in being human', *New York Times*, 2 February 2023, www.nytimes.com/2023/02/02/opinion/ai-human-education.html [accessed 6 December 2024].

14 Robert Yin, *Case Study Research and Applications: Design and Methods* (SAGE Publications, 2018), p. 15.
15 Richard Phillips and Helen Kara, *Creative Writing for Social Research: A Practical Guide* (Policy Press, 2021), p. 126.
16 Natasha Bell, 'Cruising for intellectual mothers: How writers use theory to explore the personal and the personal to explore theory', *Writing in Practice*, 4 (2022), www.nawe.co.uk/DB/current-wip-edition-2/articles/cruising-for-intellectual-mothers-how-writers-use-theory-to-explore-the-personal-and-the-personal-to-explore-theory.html [accessed 6 December 2024].
17 Graeme Harper, *Critical Approaches to Creative Writing* (Routledge, 2019), p. 3.
18 Paul John Eakin, 'Introduction: Mapping the ethics of life writing', in *The Ethics of Life Writing*, ed. by Paul John Eakin (Cornell University Press, 2004), pp. 1–18.
19 Graham Gardner, 'Unreliable memories and other contingencies: Problems with biographical knowledge', *Qualitative Research*, 1.2 (2001), pp. 185–204.
20 Charles Lemert, *Social Theory: The Multicultural and Classic Readings* (Westview Press, 1999).
21 Adapted from Phillips and Kara, *Creative Writing for Social Research*.

3

Poetry

Introduction

In Western culture, poetry is sometimes thought of as exclusive and inaccessible.[1] Yet poetry is present in our everyday lives: in greeting cards, songs, advertising slogans, as ceremonial readings, and in many other places besides. Since 1986 in London, UK, the 'Poems on the Underground' programme has displayed poems on tube trains for the pleasure of travellers.[2] Since the late 1990s the UK city of Sheffield has run a project called 'Text in the City', where poets are commissioned to create poems to be converted into pieces of public art.[3] There is an example in Figure 3.1.

Another misconception is that poets only exist in the arts. There are a number of notable scientist-poets, such as the Czech immunologist Miroslav Holub, Portuguese surgeon João Luís Barreto Guimarães,[4] and Canadian astronomer Rebecca Elson.[5] The nineteenth-century Russian mathematician Sofia Vasilyevna Kovalevskaya said, 'It is impossible to be a mathematician without being a poet in soul.'[6] Poetry appears in unexpected workplaces too, such as school chemistry classrooms,[7] training for doctors,[8] research data analysis,[9] and business strategy meetings.[10] So clearly poetry has more to offer than pure aesthetic enjoyment.

In Ethiopia, poetry is an everyday tradition and there is a long tradition of creating and sharing poetry to express feelings and thoughts about topical issues.[11] Some of these poems become popular and are widely shared, presumably because they capture a common thought, feeling, or experience. These different forms of everyday-ness might help us to think of poetry as not only the province of 'high art'.

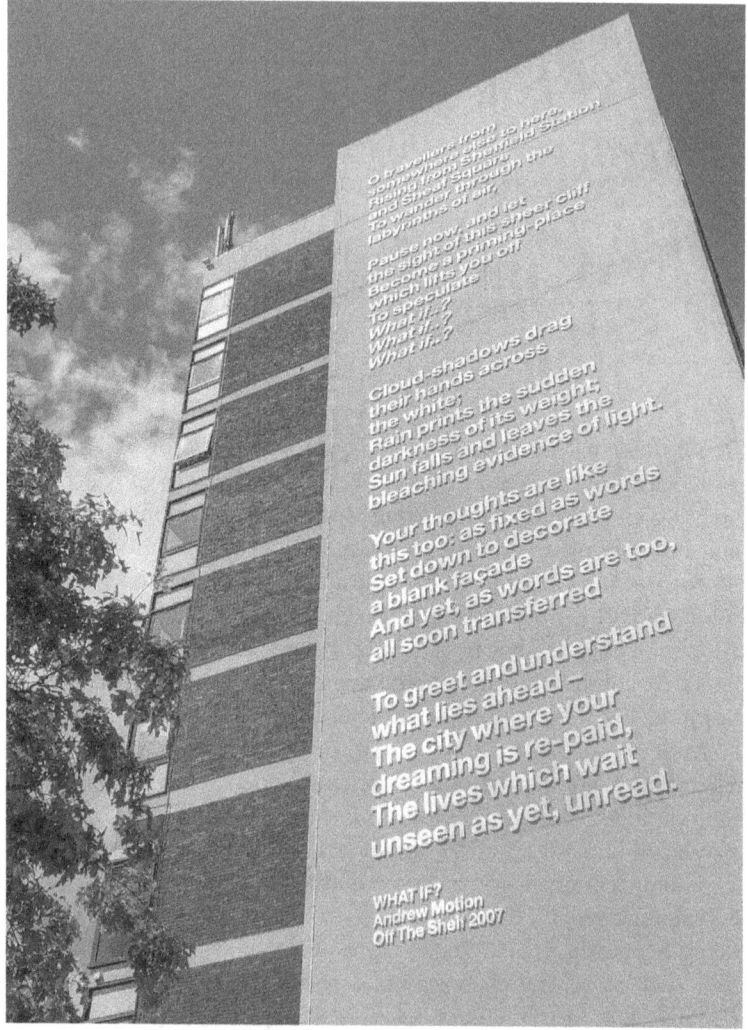

Figure 3.1 Urban poetry
Photo by Gary Butterfield on Unsplash

Poetry can help us think. Reading poetry can do this; so can listening to poetry; and, most importantly for our purposes, writing poetry can too. Contrary to the misperceptions with which this chapter started, poetry can be accessible, adaptable, and flexible.[12] To work with poetry, you don't need to know all about rhyme and metre, or to have a degree in the literary arts. You only need to be able to read, write, think, and feel. Poetry grows from our bodies: from brains and emotions and experiences, and from our bodies' rhythms such as 'the heartbeat and the pulse, the footsteps and the breath'.[13]

Professionals in fields as varied as science communication, business management, and medicine have found poetry to be a powerful and useful tool for thinking. This is because poetry privileges ambiguity and sidelines logic. Paradoxically, while poetry is made from words, it allows us to reach our 'precategorical realm' or subconscious, i.e., the part of our brain that functions without words, independently of our will.[14] Activities such as sleeping, walking, and performing mundane tasks can also help us reach this realm; we can't get there by trying, but if we do something else instead, we may be able to access this mysterious source of help inside our heads. Of course we can also use poetry for conscious thought – and indeed poetry often operates on both levels at once, or oscillates between them.

There are other paradoxes in and about poetry. A poem can be simultaneously simple and complex.[15] Partly for this reason, some people in positions of authority, from Plato onwards, have feared and distrusted poetry.[16] That is understandable because poetry can be quite subversive. When someone writes a poem they imbue their words with meaning, yet when someone reads or hears a poem, they may take a very different meaning from those words.[17] So, messages conveyed through poetry cannot be controlled, which makes poetry potentially dangerous. This, however, renders poetry fertile ground for thinking, imagining, and innovating. This is another poetic paradox.

As well as helping us to think, imagine, and innovate, poetry is also 'an extremely flexible, adaptable, and accessible medium for communication'.[18] Thinking, imagining, innovating, and communicating are essential processes in most workplaces.

What is a poem?

So what is a poem? It is a piece of writing, usually short, and definitely rhythmic.[19] The writer may use a recognised poetic form such as a limerick or sonnet, or devise a new poetic form[20] and use that, or write in 'free verse' which doesn't follow a prescribed form but 'tends to follow the rhythm of natural speech'.[21] In fact speech is much closer to poetry than to prose. The linguist Robert Carter notes that 'patterns and forms of language which as a student of literature I had readily classified as poetic or literary can be seen to be regularly occurring in everyday conversational exchanges.'[22] The poet Glyn Maxwell asserts that 'barmen talk in sestinas'[23] – a conclusion he claims to have reached on the basis of extensive research. (See below for more about sestinas.)

This explains why transcribed speech, rendered as prose, often reads oddly. Here is an example from the transcript of a keynote speech from a conference I attended on how to provide responsible and ethical scientific advice to policy-makers.

> Where to start? It's such a complex problem, but a key word that springs to my mind is trust. Trust is the basis for society to function. All relationships are based on trust. A child unconditionally trusts its mother, and when we feel unwell we go and seek help from a medical doctor whom we trust. And we need to trust the doctor to help us. Equally, research is based on trust and trusting the data that we have obtained.

The speaker here is eloquent and fluent, but even so the text reads oddly. The repetition of 'trust' helps to emphasise the point for listeners to a speech, but seems overdone in prose read from the page. Render the text as a poem, though, and it comes across quite differently.

> Where to start?
> It's such a complex problem,
> But a key word that springs to my mind
> Is trust.
> Trust is the basis for society to function.
> All relationships are based on trust.
> A child unconditionally trusts its mother
> And when we feel unwell
> We go and seek help
> From a medical doctor
> Whom we trust.
> And we need to trust the doctor to help us.
> Equally, research is based on trust
> And trusting the data that we have obtained.

Here I find that the repetition of 'trust' doesn't become wearing, as it does for me in the prose version. In the poetic version it seems a more intriguing exploration of the role of trust in society. Of course your experience of reading these pieces may be different from mine, as even two quite similar people may respond very differently to the same piece of poetry or prose.[24] But I bet the poetic version will not evoke exactly the same response as the prose version.

Poetic forms

Poetic forms are many and varied – too many to catalogue here, so I will outline some of the more common shorter forms.

Acrostic poetry

An acrostic is a poem in which the first letters of each line spell out a word or a message. They may be long or short, or not even poems at all but sentences used as mnemonics. At primary school I was taught Richard Of York Gave Battle In Vain as a way to remember the colours of the rainbow in order: red, orange, yellow, green, blue, indigo, and violet.

Here is an example of a short acrostic poem:

Craft a yummy mouthful,
Averagely nutritious.
Keeps a while if you do not
Eat the lot. Delicious!

You can see that the first letter of each line, read downwards vertically, spells out the word 'cake' – and that the poem is about a cake. An acrostic poem does not have to be about the word or message spelled out by the first letters of the lines, but if it is it will communicate more effectively. You can devise acrostics for the names of companies, departments, teams, projects, documents, key words, and so on. They can use any kind of rhyming scheme, or none. You have probably noticed that the acrostic poem above uses an ABCB rhyming scheme – more about these later – and also has an internal rhyme ('not' and 'lot').

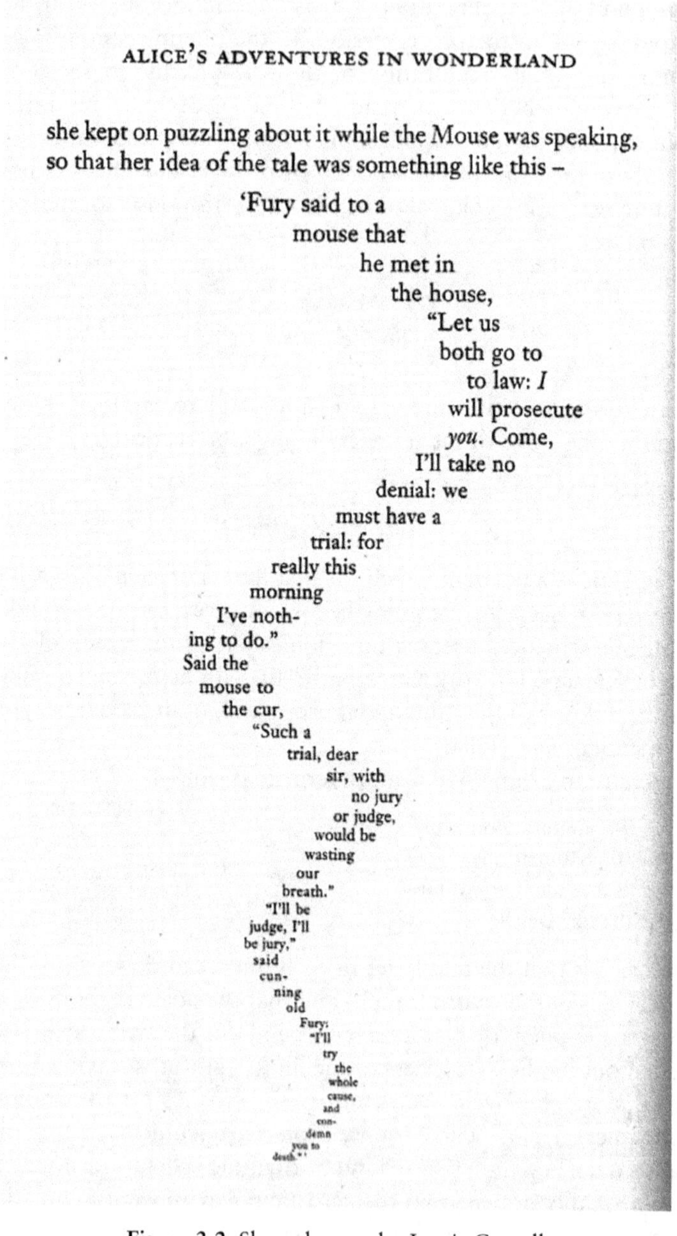

ALICE'S ADVENTURES IN WONDERLAND

she kept on puzzling about it while the Mouse was speaking, so that her idea of the tale was something like this –

'Fury said to a
mouse that
he met in
the house,
"Let us
both go to
to law: *I*
will prosecute
you. Come,
I'll take no
denial: we
must have a
trial: for
really this
morning
I've noth-
ing to do."
Said the
mouse to
the cur,
"Such a
trial, dear
sir, with
no jury
or judge,
would be
wasting
our
breath."
"I'll be
judge, I'll
be jury,"
said
cun-
ning
old
Fury:
"I'll
try
the
whole
cause,
and
con-
demn
you to
death."'

Figure 3.2 Shaped poem by Lewis Carroll

Shaped poetry

This is also known as concrete or visual poetry, and denotes a poem in the visual shape of a relevant object. A classic example is 'The mouse's tale' from *Alice in Wonderland* by Lewis Carroll, published in 1868.[25] A mouse tells Alice that his is a 'long, sad tale' but she hears 'tail', and this causes confusion:

A more recent example was created by a student on one of my creative writing courses, Gabriela Alvarado, based in the US. She wrote a short poem about abortion rights and designed it into the shape of an egg.

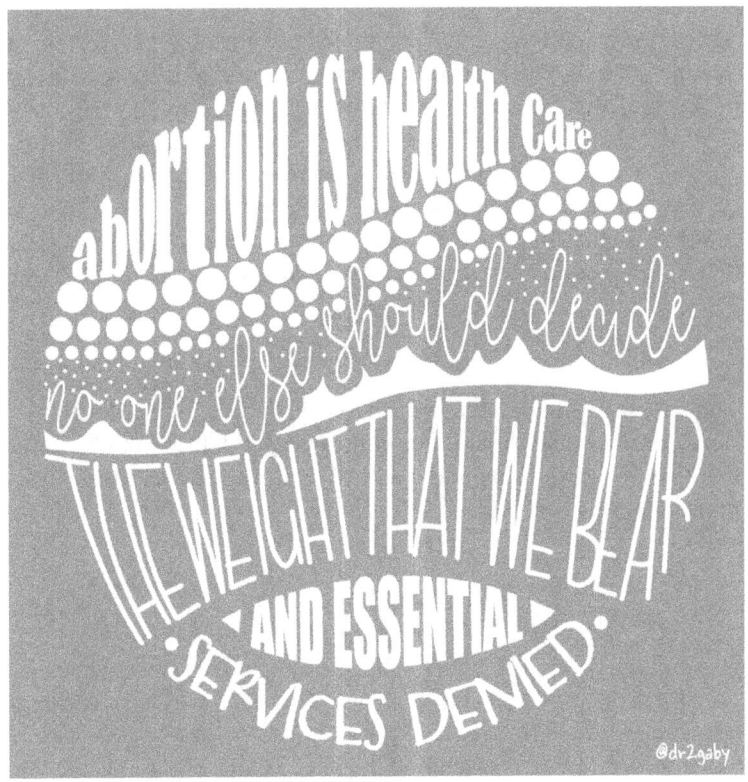

Figure 3.3 Shaped poem by Gaby Alvarado © Gaby Alvarado

Zoë John, writing about sexism in mixed martial arts, created a poem called 'Fighting Phallusy' in the shape of an erect penis. She reflects that using 'different formats can challenge or change initial ways of seeing the content'.[26]

Couplet

This is a poem with two lines, which may or may not rhyme; if they do, it is sometimes called a rhyming couplet. A couplet can be a poem in itself, as in the amusingly irreverent 'Forgive, O Lord' by the twentieth-century American poet Robert Frost:

> Forgive, O Lord, my little jokes on Thee
> And I'll forgive Thy great big one on me.

Alternatively, couplets can be combined to form longer poems, or included in longer poems such as sonnets and villanelles. Children's poems are often written in couplets, such as the mighty *How the Grinch Stole Christmas* by Dr. Seuss.[27] That is a narrative poem – a poem which tells a story – and it has over 130 lines, all of them rhyming and most of them couplets. Though the author has the sense to break up the couplets from time to time because, like any literary device, if the couplet is over-used it becomes tedious to read or hear.

Haiku

This is usually taught in English contexts as a Japanese style of short poetry which should have three lines, with five syllables in the first line, seven in the second, and five in the third. There are usually no rhymes or titles. Here is an example:

> Catty friend for life
> Always happy to see me
> Tail high in greeting

If you have spotted that this haiku is also an acrostic poem, well done! This demonstrates that sometimes it is possible to combine poetic forms.

However, the way in which haiku are taught in the English language is factually incorrect.[28] Japanese haiku are usually written in one line, not three, and that line has seventeen sound units. Sound units are not the same as syllables, and most people who speak both languages agree that seventeen sound units in Japanese equate to around ten to fourteen syllables in English.[29] Rather than counting syllables, it is more important to portray an everyday event using as few words as possible and some sensory imagery. Haiku should also prioritise nouns and verbs over adjectives, adverbs, and articles, and offer some connection to the natural world.[30]

Limerick

A limerick is a poem that is usually comic, sometimes rude, and has a prescribed rhythm and rhyme sequence. It has five lines. The first, second, and fifth are longer and they all rhyme with each other, the third and fourth are shorter and form a couplet. This rhyming scheme is often depicted as AABBA, with A referring to one set of rhymes and B to the other. Here is an example (the author is unknown):

> There was a young man from Devizes
> Whose ears were of different sizes.
> One was so small
> It was no ear at all
> The other was big and won prizes.

This is mildly amusing if you haven't come across it before. There is a rude version in which the ears are replaced with a different part of a man's anatomy; this causes much hilarity among young boys.

Quatrain

A quatrain is a four-line poem with a variety of possible rhyme schemes such as AAAA, ABAB, ABCB, and ABBA. Again, a quatrain can be a complete poem in itself, or quatrains can be combined to form longer poems. And the quatrain can be combined with some other forms: you may have spotted that the 'cake' acrostic poem, above, is also a quatrain. The twentieth-century American comic poet Ogden Nash wrote several quatrains, such as:

> To keep your marriage brimming
> With love in the loving cup
> Whenever you're wrong, admit it
> Whenever you're right, shut up.

This has an ABCB rhyme scheme as the second and fourth lines rhyme, the first and third do not.

Then there are longer, more complicated poetic forms. Examples include the sonnet, which has fourteen lines, usually in iambic pentameter; the villanelle, a nineteen-line poem with five sets of three lines and a final quatrain; and the sestina, which has seven verses, six of six lines each and a final verse of three lines, with a complex pattern of repetitions (see below for more on this). These longer and more difficult forms may be best left to experienced poets – though if you want to have a go, of course you can.

Poetry in the workplace

By this point you might be asking: What is the relevance of all this to work? In business, one classic use of poetry is in advertising slogans. These may rhyme, as with the slogan 'Once you pop, you can't stop', from the snack food Pringles (which comes in a pop-top carton). Or they may be syllabic poetry, such as: 'Maybe she's born with it. Maybe it's Maybelline.' This slogan for a make-up brand has two lines with six syllables in each. It also uses the almost musical repetition of 'Maybe ... Maybe ... Maybelline' to emphasise its message. Poetry has been used as a marketing tool in other ways too. In April 1999, around the time the new Volkswagen Beetle car went on sale, the company put forty thousand copies of a poetry book in the glove compartments of its new cars.[31] In 2000, American Airlines bought a hundred thousand copies of a poetry anthology for flight attendants to give to passengers in National Poetry Month.[32] There are many other examples too.

Poetry can also be used for internal communication in the workplace. I have already mentioned a medical research laboratory in Perth, Australia, where the lab book is kept in limericks.[33] Research has found that poetry can help employees to access and communicate tacit knowledge and so develop new insights and improve organisational learning.[34]

Sometimes people ask me why I think poetic forms are worth the bother and why we don't all just write free verse. Those are actually two separate questions. Let's start with free verse. Writing free verse is not as simple as putting words together in a vaguely poetic way and arranging them in shortish lines. A poem needs to have meaning and resonance, sound musical, and make visual sense with intelligent use of line length, line breaks, and the way words look on the page.[35] This is why poetic forms are worth the bother: because writing within the constraints of poetic forms teaches a writer how to use meaning, resonance, sound, and vision. (See below for more on constraints and creativity in writing.) Then the writer can take their learning from working with poetic forms and apply that learning to free verse.

That said, if you want to start with free verse, go ahead. There is no right or wrong way to start writing poetry.[36] I find the structures of poetic forms help me to start writing, and 'bring a welcome certainty to a process marked by uncertainty'.[37] But every writer is different, and some find the restrictions imposed by poetic forms to be more of a barrier than an enabler.

Found poems

Other barriers include fears: fears that we can't do it, that if we do produce a poem it won't be good enough, that other people will laugh at us, and so on. One poetic technique that can help overcome these fears, and has great potential in the workplace, is the 'found poem'. Found poetry is the literary equivalent of collage or remix, using existing texts in various ways to create new poems. One method is the 'cut-up poem'.[38] To create a cut-up found poem, first extract some words and phrases from an existing text, either with a theme in mind or simply by choosing the ones that appeal to you. Then rearrange those words and phrases into a poem. This may be a sequential found poem, in which the words and phrases you have chosen appear in the poem in the same order as in the source text. Or it may be a non-sequential found poem, in which you choose the order of the words and phrases in your poem.

Another method of constructing a found poem is the 'blackout poem', also known as the 'erasure poem', where most of the words in a page or section of text are blacked out and the remaining words constitute a poem. Here is a blackout poem I have created from the executive summary of a UK government policy document on transport in rural areas.[39] I am not claiming that this is a particularly

Executive summary

Innovation in transport technologies and services has the potential to enhance rural transport and support a higher quality of life for people in rural areas.

Figure 3.4 Blackout poem

'people in rural areas / unable / isolated, cut off / out of reach / enhance quality of life / preserve greater choice / demand services, connectivity / unlock rural areas / tackle best practice.'

good poem but, like the best blackout poems, it is in dialogue with its source.[40]

Found poems can be created quite quickly. Creating found poems 'challenges you to construct and extract new meanings from the original text, which is a great exercise in creativity and critical thinking'.[41] This means that, with suitable texts to work from, creating found poems can be a useful exercise in business strategy meetings, teaching, data analysis, staff awaydays, training, and a host of other professional tasks and events. 'Suitable texts' can include law reports, advertisements, newspaper articles, even street signs or other poems[42] – anything that is relevant to your work.

Another type of found poem is the I-poem. This technique was devised and developed by researchers in the late twentieth and early twenty-first centuries. It is useful for working with any text written in the first person: a transcript of a meeting, a novel, an appraisal form, and so on. An I-poem is constructed by finding and extracting each statement that begins with 'I' or where 'I' plays a prominent role. Then each statement is placed on a new line, in the same order as they appeared in the source text, to construct the I-poem.

This process is more straightforward in theory than in practice. A statement beginning with I may be short or long; if it is long, do you include the whole thing? If not, where do you stop? What constitutes 'a prominent role'? What do you do when there is a statement you really want to include which doesn't contain 'I'? There are many decisions like these to be made in constructing an I-poem.

An I-poem distils 'the subjective experience of the storyteller' and enables us to understand more fully the complexity of, and contradictions within, their experience.[43] To show how this can work in practice, here is an example I have constructed from an article by the Korean illustrator Henn Kim in the Guardian's *A Moment That Changed Me* series:[44]

> I was mute for two years
> I couldn't convey my emotions
> I felt that the everyday noises around me were like a battlefield
> When I was 17, I started wearing headphones
> I was 19 when I decided to stop speaking
> I felt trapped because I couldn't express my emotions
> Without words, I felt inspired
> I became obsessed with music

I loved the Smiths, David Bowie and Björk
I listened to it on repeat
I became deeply introspective
Sometimes, I felt lethargic and couldn't get out of bed
I watched films non-stop
I distanced myself from people
I started to communicate via text messages
I moved to Seoul
There I met like-minded people who loved art
Finally, I began to speak a little
I'm still afraid of speaking
I try not to reveal too much about myself
I have a new career as an illustrator
I was mute
I witnessed the power of art
I still prefer texting over speaking

The source text is richly eloquent, with many more words and beautiful illustrations which complement the writing. The I-poem above could be seen as a summary of the article, but it is not, it is a distillation of Henn Kim's late teenage years. So I-poems are a way to focus in on one person's experience.

The same process can be used to construct we-poems, they-poems, and so on. We-poems can help us to understand people's relationships and sense of belonging.[45] Suitable source texts include organisational documents such as team meeting minutes or annual reports. They-poems can be useful in understanding how people, groups, and organisations position other people, groups, and organisations.[46] Suitable source texts include diaries and policy documents.

So far we have covered single-pronoun poems, but of course you could focus on more than one personal pronoun at a time. We-poems can often work well as we-and-us-poems.[47] I-poems can work well as I-and-we-poems, extending the focus on experience to include relationships. Pronoun poems, where all personal pronouns are included, can teach us about, and convey, the complexities and contradictions in people's experiences, relationships, and positioning of other people and groups.

A natural development from this is the keyword poem, invented by me. This is a poem focusing on any aspect of the world that can be encapsulated in a single word or a set of related words. So for a keyword poem on motherhood, you could include also mother, mum/mom/momma/mommy/mamma/ma, maternal, and so on. Here is an example of a keyword poem I constructed from a newspaper article about the Japanese tennis player Naomi Osaka's experience of motherhood.[48]

> Motherhood has changed her perspective on tennis and the world
> Becoming a mother has helped her mental health
> Approaching the tournament as a mother
> It seems so far apart from being a mom
> I often worry about if I'm a good mom

This short poem clearly demonstrates the complexity and contradictions that can be highlighted by these kinds of found poems. The first three-line verse is in the journalist's voice, telling us about some positive and upbeat aspects of an elite sportswoman's experience of motherhood. Then the second two-line verse is in Osaka's own voice which conveys uncertainty and anxiety. This contrast would be difficult to discern from the article itself, in which the quote from Osaka is placed at the end, like an afterthought. Also, in that quote, she says several other things as well as her two direct mentions of motherhood.

Constraints and creativity

Poetry, perhaps more than any other style of writing, uses constraints to promote creativity. Free verse is comparatively unconstrained but that does not mean it is easier to write – or, at least, it is not easier to write well. All the poetic forms, and the techniques of found poetry, impose constraints that creativity can, paradoxically, thrive within. Glyn Maxwell reports the twentieth-century British-American poet W. H. Auden's view that these constraints 'forbid automatic responses' and 'force us to have second thoughts'.[49] Our first thoughts are often involuntary, while our second thoughts are more considered, which means they are often more interesting and useful than our first thoughts.

Writing is always a dance between invention and construction. Words are our building blocks and we have innumerable choices about how to put them together to express the points we want to make. In personal writing, such as a personal diary or journal, we are free to write whatever and however we like. In workplace writing there are more constraints – yet, as we have seen, constraints can enable and support creativity.

This is evident in the solving of word puzzles which requires creative thought as well as trial and error.[50] This applies to simple puzzles such as 'how many words can you make from the letters in "outback"?' as much as to complex puzzles such as Scrabblegrams. A Scrabblegram is a poem or story created from all one hundred tiles in an English Scrabble game. One of the best examples was written by David Cohen in 1997, and reads:

> A clown jumps above a trapeze
> Arcs over one-eighty degrees
> Out into midair
> Quite unaware
> Of his exiting billfold and keys[51]

This is particularly clever because it also forms a perfect limerick.

Writing poems can, at times, be a bit like doing word puzzles. If you want to create an acrostic poem, you are constrained by the word or message spelled out from the first letter of each line. Writing the other comparatively simple poetic forms described above is constrained by numbers of lines and/or rhyming schemes.

Some poetic forms are much more complex and, as a result, impose more constraints. The sestina is a poetic form, dating from the twelfth century, which has a strict structure with six verses of six lines each and a seventh verse of three lines. There is also a specific pattern of repetition and sometimes rhyme. The last words of the six lines in the first verse are also the last words of the six lines in the second, third, fourth, fifth, and sixth verses, but in a rotating order which is different in each verse.[52] So if we use ABCDEF to represent the last words of the six lines in the first verse, the last words of the second verse could be FAEBDC, in the third they could be CFDABE, in the fourth ECBFAD, in the fifth DEACFB, and in the sixth BDFECA. In the final three-line verse, all six words must be used again, with three at the ends of the lines and three contained within the lines. This is done in various ways, such as (A)B, (C)D and (E)F or (B)E, (D)C, and (A)F. Rhymes may also be included such that, in the first verse, lines A, C, and E rhyme, and so do lines B, D, and F.

This means that, when you have written the first verse of a sestina, you know the last word of each line for the rest of the poem.[53]

So, creating the rest of the sestina is a bit like solving a word puzzle. Some sestinas read like word puzzles; others (the cleverer ones, I think) read like poems. Here is one by contemporary American poet Joan Larkin:

Jewish Food[54]

It's the worst – but it tastes so good. (Gerald Stern)

I came from school to warm bread
and *tsibele bulkes*: Russian rolls, onions
in wells her palm pressed in the dough, softened in sweet
butter and baked in. Little pillows, fragrant as flesh.
I'd eat a few with cold milk at two. Five-thirty, supper
was on the table, Dad home between shows and hungry

for soup with *knaidlach* and boiled chicken. I was still hungry
afterward for a heel of black bread
smeared with rendered chicken fat. *Shabbes*, supper
had to be chicken. No milk on the table. Onions,
salt and fat were what she put in chopped liver, start of a *fleyshedik* meal. To end it, fruit in thin, sweet

syrup: compote – pears, prunes plumped with cooking, sweet-
ened with raisins. No one left the table hungry
or thought there was anything wrong with fattening childflesh
at three meals and between: *mon* cookies, hunks of rye bread,
batter licked from the bowl. I watched her knife cleave onions,
carrots for *tzimmes*, beets for borscht. *Pesach*, supper

was called a *Seder* – not an ordinary supper.
Matzo folded in a cloth napkin, goblets filled with sweet
red wine – they spilled drops for each of the plagues. Glazed onions
and brisket waited while uncles prayed. I sat there hungry,
wondering at the strangeness of a week with no bread.
In candlelight, my grandmother's warm flesh-

folds shone, the rough crepe of her peasant flesh
smoothed with Jergens lotion. She scoured sinkfuls of pots after supper
then sat and ate some of her own unleavened bread
baked with *matzo mel* and sucked sweet
tea through cubes of sugar. I sat with her, hungry
for stories of the old days, when sometimes even onions

were scarce but everyone told jokes. Onions
couldn't make you cry if you ran water while you cut their raw flesh.

She always knew you were hungry, everyone was hungry,
and she sneaked cream into your coffee, if it was a *milchik* supper,
even if you said you wanted it black. Her voice was Russian music, sweet
even when she said harsh things. I can't think of her without tasting bread –

no one made better bread. She gave me the taste for onions,
the oily flesh of a carp, the cold thick sweet-
ness of sour cream on a blintz for supper. God forbid I should be hungry.

Tsibele bulkes: onion rolls
Knaidlach: matzo balls
Shabbes: the Sabbath
Fleyshedik: made with meat
Mon: poppy seed
Pesach: Passover
Seder: ritual Passover meal
Matzo: unleavened bread
Matzo mel: flour made from ground matzo
Milchik: made with dairy products

Writing sestinas is not for everyone. I have never tried to create a sestina. I include them here more for inspiration than for emulation – though if you want to try your hand at crafting a sestina, then give it a go.

Another poetic paradox is that the constraints of poetry render it useful in some contexts which are more usually associated with methodical, systematic work. In academia, some researchers write poetry to help them analyse their data. This may be found poetry constructed from textual data[55] or poetry written by a researcher to help them reflect on their data (of any kind) and its meanings.[56]

Conclusion

As we saw in Chapter 1, writing is a very emotional process, and writing poetry is no exception. Even the thought of writing a poem may provoke thrilling anticipation in some people and deep dread in others. Whatever your own feelings may be, it can be hard to figure out where to start: choose a form, pick a method, or just write some words? It may help to know that there is no 'right way' to start work on a poem, and the best way to start is exactly that: start![57]

One of the most useful qualities of poetry is its ability to get to the core, the essence, of a topic.[58] That topic could be anything from dramatic to mundane or from personal to professional. Poetry can achieve a lot in a short space, almost like an illustration made from words. It also emphasises ambiguity, which can help people at work to develop the skills necessary to cope with complexity and rapid change.[59] The evidence on which this chapter is based makes it clear that reading and writing poetry can help us to understand, strategise,[60] analyse, and communicate.[61] Those are common workplace requirements so it makes sense to use poetry, among other techniques, to assist with these functions in our working lives.

Sending up a fibro flare

My hips
Are made of paperclips.
My hands won't grip.
Joints crack; I trip.
I need to limp
On both feet
Which can't be done
Or, as the French say,
Il ne marche pas.
This flare is no damn fun.
But still, I can appreciate a good bilingual pun.

Try it yourself

You have two options to choose from:

Option 1. Write a poem about an aspect of your work or workplace. The length is up to you, as is whether you use free verse or a poetic form. You can even devise a new form if you like. Give your poem a little time and care; try to make it as good as you can. Don't stress about it though: remember that this exercise is only for your own purposes, and there is no need to show anyone the finished product unless you want to. If this prospect feels too daunting, try the next option.

Option 2. Create a found poem about an aspect of your work from a suitable source: perhaps a workplace document, or a document you find online about the type of work you do, or your own reflective journal. Your found poem may be sequential or non-sequential, or an I-poem or other pronoun poem, or a blackout/erasure poem – or, again, you can devise a new form of found poetry, such as by using the seventh word of each sentence. Take time to think about the source and the work you are doing to create your poem from that source.

Whichever option you choose, when you have finished, give yourself a few minutes to reflect on your poem. Here are some reflective questions that may help.

- How did you find the experience of writing/creating your poem?
- Did you learn anything from the process?
- Did anything surprise you?
- Would you use this method again in the future? Why?

Notes

1 Melisa Cahnmann-Taylor, 'The craft, practice, and possibility of poetry in educational research', in *Poetic Inquiry: Vibrant Voices in the Social Sciences*, ed. by Monica Prendergast, Carl Leggo, and Pauline Sameshima (Sense Publishing, 2009), pp. 13–29.
2 'About us', Poems on the Underground, https://poemsontheunderground.org/about-us [accessed 19 January 2024].
3 Julia Armstrong, 'Sheffield writer Warda Yassin unveils love poem for city in Orchard Square', *The Star*, 2 November 2021, www.thestar.co.uk/whats-on/arts-and-entertainment/writer-unveils-love-poem-for-city-in-orchard-square-3442593 [accessed 19 January 2024].
4 Oliver Balch, '"As with a poem, each patient is unique": The cancer surgeon using poetry to help train doctors', *The Guardian*, 17 February 2024, www.theguardian.com/science/2024/feb/17/joao-luis-barreto-guimaraes-cancer-surgeon-poetry-pessoa-prize [accessed 18 February 2024].
5 Sam Illingworth, 'Scientists and poets are more alike than you might think', The Conversation, 31 May 2019, https://theconversation.com/scientists-and-poets-are-more-alike-than-you-might-think-116326 [accessed 18 February 2024].

6 'Sofia Kovalevskaya: The girl who wouldn't give up on math', Mathnasium, 22 January 2018, www.mathnasium.com/math-centers/littleton/news/sofia-kovalevskaya-the-girl-who-wouldnt-give-up-on-math-783324179 [accessed 23 August 2024].

7 João Carlos Paiva, Carla Morais, and Luciano Moreira, 'Specialization, chemistry, and poetry: Challenging chemistry boundaries', *Journal of Chemical Education*, 90 (2013), pp. 1577–1579 (p. 1578), https://doi.org/10.1021/ed40030891

8 Debbie McCulliss, 'Poetic inquiry and multidisciplinary qualitative research', *Journal of Poetry Therapy*, 26.2 (2013), pp. 83–114 (p. 98).

9 Richard Phillips and Helen Kara, *Creative Writing for Social Research: A Practical Guide* (Policy Press, 2021), pp. 127–132.

10 Clare Morgan, Kirsten Lange, and Ted Buswick, *What Poetry Brings to Business* (University of Michigan Press, 2010), pp. 59–62.

11 Setargew Kenaw, 'Cultural translation of technologies in Ethiopia', *Ethiopian Journal of the Social Sciences and Humanities*, 13.2 (2017), pp. 109–142 (p. 121), https://doi.org/10.4314/ejossah.v13i2.5

12 Sam Illingworth, *Science Communication through Poetry* (Springer Nature, 2022), p. 3.

13 Gavin Maxwell, *On Poetry* (Bloomsbury Academic, 2012), p. 80.

14 Morgan, Lange, and Buswick, *What Poetry Brings to Business*, p. 52.

15 Jonathan Davidson, *On Poetry* (Smith|Doorstop Books, 2018), p. 7.

16 Ben Lerner, 'From "The hatred of poetry"', *Poetry*, 1 April 2016, www.poetryfoundation.org/articles/88730/from-the-hatred-of-poetry [accessed 10 January 2024].

17 Kimberley Dark, 'Examining praise from the audience', in *Poetic Inquiry: Vibrant Voices in the Social Sciences*, ed. by Monica Prendergast, Carl Leggo, and Pauline Sameshima (Sense Publishing, 2009), pp. 171–185 (p. 175).

18 Illingworth, *Science Communication through Poetry*, p. 3.

19 *Ibid.*, p. 30.

20 Stephen Fry, *The Ode Less Travelled: Unlocking the Poet Within* (Arrow Books, 2005), p. xviii.

21 Illingworth, *Science Communication through Poetry*, p. 42.

22 Ronald Carter, *Language and Creativity: The Art of Common Talk* (Routledge, 2004), p. 10.

23 Maxwell, *On Poetry*, p. 62.

24 Illingworth, *Science Communication through Poetry*, p. 14.

25 Andrew Sellon, 'Alice discovers concrete (poetry) in Wonderland!', Lewis Carroll Society of North America, 31 May 2013, www.lewiscarroll.org/2013/05/31/alice-discovers-concrete-poetry-in-wonderland/ [accessed 9 October 2023].

26 Zoë John, 'Grappling with poetry: Why to start and how to start', in *Handbook of Creative Research Methods*, ed. by Helen Kara (Bloomsbury Academic, 2024), pp. 95–106 (p. 99).
27 Dr. Seuss, *How the Grinch stole Christmas*, Best Poems Encyclopedia, www.best-poems.net/poem/how-grinch-stole-christmas-by-dr-seuss.html [accessed 1 January 2024].
28 Kirsten Deane, Raphael d'Abdon, and Angela Hough, 'The way of poems', in *Poetic Inquiry as Research: A Decolonial Guide*, ed. by Heidi van Rooyen and Raphael d'Abdon (Policy Press, 2025 [in press]).
29 *Ibid.*
30 *Ibid.*
31 Robyn Meredith, 'The media business: Advertising; marketing departments are turning to poets to help inspire their companies' clientele', *New York Times*, 21 March 2000, www.nytimes.com/2000/03/21/business/media-business-advertising-marketing-departments-are-turning-poets-help-inspire.html [accessed 6 December 2024].
32 *Ibid.*
33 Jillian Swaine (2017), personal communication.
34 Louise Grisoni, 'Cooking up a storm: Flavouring organisational learning with poetry', paper presented at the International Conference on Organisational Learning, Knowledge and Capabilities (OLKC) (London, Ontario, Canada, June 2007), https://warwick.ac.uk/fac/soc/wbs/conf/olkc/archive/olkc2/papers/grisoni.pdf [accessed 6 December 2024].
35 Maxwell, *On Poetry*, pp. 24–25.
36 Zoë John, 'Grappling with poetry: why to start and how to start', in *The Bloomsbury Handbook of Creative Research Methods*, ed. by Helen Kara (Bloomsbury, 2023), pp. 95–106 (p. 104).
37 May Huang, 'On sestinas and literary translation', *Words without Borders*, 3 February 2021, https://wordswithoutborders.org/read/article/2021-02/on-sestinas-and-literary-translation-may-huang/ [accessed 16 February 2024].
38 Kedrick James, 'Cut-up consciousness and talking trash: poetic inquiry and the spambot's text', in *Poetic Inquiry: Vibrant Voices in the Social Sciences*, ed. by Monica Prendergast, Carl Leggo, and Pauline Sameshima (Sense Publishing, 2009), pp. 59–74.
39 Department for Transport, *Future of Transport: Helping Local Authorities to Unlock the Benefits of Technology and Innovation in Rural Transport* (Department for Transport, 2023), https://assets.publishing.service.gov.uk/media/651c266bbef21800156decb0/future-of-transport-helping-local-authorities-to-unlock-the-benefits-of-technology-and-innovation-in-rural-transport.pdf [accessed 8 October 2023].

40 Sean Glatch, 'What is blackout poetry? Examples and inspiration', Writers.com, 2 October 2023, https://writers.com/what-is-blackout-poetry-examples-and-inspiration [accessed 8 October 2023].

41 Kaelyn Barron, 'What is found poetry? Examples and tips for making your own found poem', TCK, www.tckpublishing.com/found-poetry/ [accessed 10 January 2024].

42 Ibid.

43 Lori Koelsch, 'The use of I poems to better understand complex subjectivities', in *Poetic Inquiry II: Seeing, Caring, Understanding*, ed. by Kathleen Galvin and Monica Prendergast (Sense Publishers, 2016), pp. 169–179 (pp. 177–178).

44 Henn Kim, 'A moment that changed me: I stopped speaking at 19 – and found my artistic voice', *The Guardian*, 13 December 2023, www.theguardian.com/lifeandstyle/2023/dec/13/a-moment-that-changed-me-i-stopped-speaking-at-19-and-found-my-artistic-voice [accessed 10 January 2024].

45 Mark Oliver Llangco, 'We-poem as sociological poetry and method of data analysis', *Philippine Sociological Review* (2019), pp. 63–96 (p. 63), www.jstor.org/stable/10.2307/26933205

46 Rachel Helme, 'I and THEY poetic voices in learning to listen to a student labelled as low attaining in mathematics', in *Proceedings of the 45th Conference of the International Group for the Psychology of Mathematics Education*, 2, ed. by Ceneida Fernández, Salvador Llinares, Angel Gutiérrez, and Núria Planas (Servicio de Publicaciones, 2022), pp. 371–378 (p. 377).

47 Gemma McKenzie, 'Freebirthing in the United Kingdom: the voice centered relational method and the (de) construction of the I-Poem', *International Journal of Qualitative Methods*, 20 (2021), pp. 1–13 (p. 10), https://doi.org/10.1177/1609406921993285

48 Naomi Osaka, 'Naomi Osaka was "shocked" by lack of paid maternity leave in US', *The Guardian*, 10 January 2024, www.theguardian.com/sport/2024/jan/10/naomi-osaka-was-shocked-by-lack-of-paid-maternity-leave-in-us [accessed 10 January 2024].

49 Maxwell, *On Poetry*, p. 113.

50 Benedict Carey, 'Tracing the spark of creative problem-solving', *New York Times*, 6 December 2010, www.nytimes.com/2010/12/07/science/07brain.html [accessed 22 January 2024].

51 Alex Bellos, 'Can you solve it? The greatest wordplay puzzle of all time', *The Guardian*, 22 January 2024, www.theguardian.com/science/2024/jan/22/can-you-solve-it-the-greatest-wordplay-puzzle-of-all-time [accessed 22 January 2024].

52 Daniel Nester, 'Introduction', in *The incredible sestina anthology*, ed. by Daniel Nester (Write Bloody Publishing, 2013), pp. 19–23 (p. 19).

53 Mark Miller (2024), personal communication.
54 Joan Larkin, 'Jewish food', *McSweeney's*, 13 April 2006, www.mcsweeneys.net/articles/jewish-food [accessed 16 February 2024]. Reproduced here with the kind permission of the poet.
55 Laurel Richardson, *Fields of Play* (Rutgers University Press, 1997), p. 135.
56 Adrie Kusserow, 'Anthropoetry', in *Crumpled Paper Boat: Experiments in Ethnographic Writing*, ed. by Anand Pandian and Stuart McLean (Duke University Press, 2017), pp. 71–90 (p. 85).
57 John, *Grappling with Poetry*, p. 104.
58 Rich Furman, 'The mundane, the existential, and the poetic', in *Poetry as Therapy, Research, and Education: Selected Works of Rich Furman*, ed. by Rich Furman (University Professors' Press, 2022), pp. 46–65 (p. 46).
59 Morgan, Lange, and Buswick, *What Poetry Brings to Business*, p. 17.
60 *Ibid.*, pp. 41–42.
61 Rich Furman, 'Beyond the literary uses of poetry: A class for university freshmen', in *Poetry as Therapy*, ed. by Furman, pp. 241–248 (p. 242).

4

Graphic writing

Introduction

Graphic writing refers to outputs such as cartoons, comics, graphic novels, and zines. The boundary between drawing and writing is not firm but blurred. Egyptian hieroglyphs were a system of written communication that included a lot of pictorial elements. Calligraphy turns writing into an art form with every letter or kanji a beautiful shape. Māori tattoos are not purely decorative, they have functions including communicating information about the wearer to others.[1] In the Vanuatu archipelago of the south Pacific, drawing always goes together with talk, and is in the form of geometric patterns drawn with a finger in sand or dust or ash. This system is used for various purposes, such as storytelling, discussion, and teaching, and the drawings are not intended to be permanent but to 'be blown away in the breeze' like the spoken words.[2]

This blurred boundary becomes less surprising when you consider that writing is a visual representation of speech. Some particularly artistic types of writing can even have spiritual significance, such as Islamic calligraphy[3] and Māori tattoos.[4] So combining writing with drawing to convey human ideas and experiences makes a lot of sense.

You may think of cartoons, comics, graphic novels, and zines as art forms rather than writing, but really they are vehicles for storytelling. Comics and their relations are made from sequential art, 'a unique and powerful form of communication'[5] which has been used to tell stories around the world for millennia. Sequential art has been found in prehistoric European cave paintings,[6] ancient

Egyptian tomb artworks, pre-Columbian scrolls and the Bayeux tapestry,[7] to name just a few. These days, sequential art is used in documents for the public, such as flight safety information cards and instructions for flat pack furniture.[8]

Cartoons

Even a single-panel cartoon can tell a story. From the Victorian era, the cartoonist Marie Duval pioneered comedic visual storytelling in the cartoons she produced for cheap newspapers.[9] Her cartoon about the late nineteenth-century craze for roller-skating rinks shows a fashionable woman wearing roller skates with high-heeled boots, her hands wrapped warmly in a muff, looking poised and composed as she skates. In the background are various people falling off their skates, and below is an assortment of crutches, bandages, doctor's bills, and legal documents. The cartoon's title describes the roller-skating craze as 'a passing fancy', and adds, in brackets, 'and the sooner it has passed away altogether the better'. This cartoon tells a story in itself: the story of social enthusiasm for a fashionable but risky activity. There is also a sub-plot to this story, as the main character and another skilful skater prominently outlined in the background are both women, while most of those falling over are men. This makes a subtle point about feminism in a humorous way. A carefully constructed cartoon is an excellent vehicle for this kind of communication – although cartoons can also enable dramatic and exaggerated point-making.

Cartoons are often engaging because they use humour, are quick to read, and appeal to visual and verbal thinkers alike. They can be created by hand or digitally, and although artistic skills will help, they are not essential. In the workplace, cartoons can be useful for all kinds of purposes, such as to liven up a PowerPoint presentation, disseminate corporate information, or illustrate a report. If you feel the need for some support in getting going with cartoon creation, there is lots of information online about how to create cartoons, as well as courses you can attend. Key points to remember are that a cartoon should be relevant and useful for your audience, and any humour used should be non-discriminatory.

Figure 4.1 Cartoon by Marie Duval

Comics in the workplace

In some English-speaking countries, sequential art-based formats are regarded as only appropriate for children and young people.[10] In some other countries, though, such as France and Belgium, comics are recognised as mainstream literary art forms.[11] In 1992, the graphic novel *Maus* by Art Spiegelman won a Pulitzer for literature,[12] and in 2018 a graphic novel, *Sabrina* by Nick Drnaso, was longlisted for the Booker prize which is another prestigious literary award.[13] So even where comics are not recognised as mainstream literary art forms, their value may still be acknowledged.

The term 'graphic novel' was coined by renowned American comics creator Will Eisner in 1978[14] to try to make the format seem more grown-up and acceptable. However, creators usually use the term 'comics', and I will follow suit. Comics can be useful in the workplace. Research in a professional setting has shown that comics improve adults' comprehension.[15] Research also tells us that using pictures and text together helps readers to remember more of the information conveyed,[16] and we know that stories also aid comprehension and memory. So, if we are writing about something we would like people to understand and remember, a sequential art-based format is a very good choice.

People in a wide range of professions have recognised this for almost a century. In 1933 a comic called *Funnies on Parade* was offered free of charge to people who bought toiletries made by Procter & Gamble, if they sent in a coupon as evidence of their purchase.[17] This proved to be a successful way of promoting products and other manufacturers began to follow suit.

During World War Two, Will Eisner – then a warrant officer in the Pentagon in Washington, DC – first used cartoons for instructional purposes in *Army Motors* magazine.[18] This approach worked so well that, after leaving the army, Eisner started a corporation to create instructional comics for the statutory and private sectors in the US.[19]

From the mid-twentieth century, comics were increasingly used in education, first in schools and more recently in colleges and universities.[20] Researchers have begun to use comics to present their findings, in fields from post-disaster ethnography[21] to parasitology.[22] In the late twentieth century, journalists around the world began to use comics to present some items of news.[23] Comics journalism is now a recognised genre and a powerful medium for conveying social commentary.[24]

The term 'graphic medicine' was first used in 2007 by the British medical doctor and comics creator, Ian Williams, to name the scope for comics to support healthcare. This is broad, including such things as: educational comics for patients, students, and healthcare professionals; memoirs of illness created using sequential art; and therapeutic workshops which involve making comics.[25] In 2019, a non-profit organisation, the Graphic Medicine International Collective, was formed 'to guide and support the use of comics in health'.[26]

Tech companies, too, have understood the benefits of comics. We saw in Chapter 1 of this book that in 2017, Apple published their iTunes terms and conditions in comic form, thereby converting dry and difficult prose into an entertaining read. When the web browser Google Chrome was launched in 2008, Google commissioned Scott McCloud, another eminent comics creator, to create a comic book explaining how the browser worked for journalists and bloggers.[27] 'It seems as though today anything can be adapted and translated into the sequential art form.'[28]

So you can see that comics have been, and are being, used in a lot of different workplaces. There are three main reasons for this. First, comics can communicate the reality of places most readers could never actually go, such as the inside of a hurricane-damaged home,[29] and experiences they are unlikely ever to have, such as living with an obscure medical syndrome.[30] This enables comics to offer insightful information in an accessible and entertaining way. Second, when comics are created well, they can have a big impact on their readers through engaging both brain hemispheres at once[31] which makes the messages they convey more memorable.[32] Third, because they use both words and pictures, comics are better at conveying complexity and nuance than either words or pictures alone.[33] Altogether, this means that comics can help to bridge the gap between theory and practice.[34] Communicating information in comprehensible and memorable ways, and bridging communication gaps, are essential functions in the world of work. This explains why sequential arts-based formats can be such a good choice.

Moreover, comics are an excellent way to help people understand complexity.[35] This means comics are useful in a range of professional activities such as advertising, coaching, conveying information, facilitation, marketing, presentations, teaching, and training. In advertising, the global confectionery company Mars used cartoon characters to advertise M&Ms. Each character was based on one of the colourful round sweets, and the campaign ran from the mid-1980s to the early 2020s.[36] I would have liked to include an image here but was not able to do so for copyright reasons, though you can find them easily on the internet. The characters became known as 'spokescandies' and were also used in animations, for showing on TV and in cinemas; many of these can still be seen online.

In coaching, comics can be used to provide examples for discussion or inspiration. There are comics on just about every subject you can think of: art and education,[37] history and philosophy,[38] politics and power,[39] the impact of colonisation on Indigenous peoples,[40] a young woman's experience of depression,[41] and many more. The ability of comics to immerse the reader in a world different from their own can be very helpful for a coach or a teacher working to develop someone's skills and abilities.

Comics are used in various settings to convey information to groups of people. The Scottish Centre for Comics Studies at the University of Dundee works with local, national, and international government departments, non-profit organisations, and others to create public information comics. These are 'on themes such as healthcare, education, human rights, science communication and sustainability'.[42]

When you are facilitating a group session, comics can help to energise people and promote creativity. You can use comics to support ice-breaking, idea-generating, or team-building activities. These may be comics that exist already, or – if the group is willing and able – comics they create in the session. These may be done very simply, such as by using Post-It notes in place of panels, to make it easy to re-draw and re-order the comic as it is produced.[43]

Comics are useful in marketing because they are attention-grabbing and memorable. Making comics visually attractive and humorous, with an engaging story, enables organisations to connect with their clients or customers in an appealing way. Comics can be used in print and digital materials, from reports to websites and social media. This flexibility is an asset to any marketing campaign.

In presentations, comics can add visual appeal, enhance the storytelling on which all presentation is based, and inject humour to make the content more memorable. You can use comics as visual aids or as handouts, or both. Presentations in comic form are often very effective in conveying information in an accessible and engaging way.

The value of comics as a teaching aid, from the very first year of school to postgraduate education and beyond, is now well known and understood. A recent review of relevant research identified a range of benefits from using comics in educational settings. These included: increased participation and motivation in class; easier learning and more retention of what had been learned; development of skills; and better academic performance.[44] As some comics experts have said:

> It is no longer a question of whether sequential art should be used in educational settings, but rather how to use it and for what purpose.[45]

Explaining how to use sequential art in educational settings would take a whole book in itself. In brief, options include: using existing comics, commissioning one or more new comics, and creating comics in the classroom: either individually, in groups, or as a whole class.

Beyond the classroom, workplace training is a key educational setting. Comics have proved useful in training for diverse personnel including airline crew,[46] qualitative researchers,[47] and senior naval submariners.[48] One reason comics are so valuable as a training aid is that, as we have seen, the format is particularly good at helping people to bridge the gap between theory and reality or practice.[49]

No doubt there are also other ways in which comics can be useful in professional life, some of which may not yet have been devised. Even so, this section shows that comics are unquestionably an asset to the workplace.

How to make comics

In comics, words and images work together to tell a story, which means comic making involves both writing and drawing. However, you don't have to be able to draw and write to make comics. If you can write but not draw, you can use a digital comics maker such as

Canva or Pixton, or you can hire an artist to work with you. If you can draw but not write, you could use an AI text generator to help produce the first draft, and/or hire a writer to work with you.

Comic making usually starts with writing. I write comics but I don't draw them. My script for *Conversation with a Purpose* began like this:

Table 4.1 *Conversation with a Purpose* by Helen Kara

Panels	Visuals	Text
Page 1	3 FPW	
1 FPW	Young adult students waiting for teacher in classroom; casual attitudes: tilting back on chairs, sitting on desks, etc. A teaching board reads INTERVIEWING in big letters. A clock on the wall says five to ten.	Three speech bubbles: I had an interview to come here. It was awful. Are research interviews like that? Like interrogation under torture?
2 FPW	Amal coming in through door, students hurrying to take their seats. Clock says one minute to ten.	Three speech bubbles: Are they going to teach us to be torturers? I hope not! Owen: It's just a conversation, isn't it?
3 FPW	Amal standing in front of seated class. Owen giving an 'I told you so' look to his friend. Clock says two minutes past ten.	Amal: Essentially, an interview is a conversation.
Page 2	2 HPW, 1 FPW, 2 HPW	
4 HPW	Close-up on the teacher.	Teacher: But it's a conversation with a purpose: to glean information. To find stuff out.

Graphic writing

Panels	Visuals	Text
5 HPW	Teacher.	Teacher: You know the basics from your reading. Now it's time to practice. Get into pairs and ask each other about a childhood time when you learned something useful outside of school.
6 FPW	Students sitting in pairs, one of each pair talking and the other listening intently. Some people serious, some smiling, some gesticulating, some sitting still.	
7 HPW	Owen and friend sitting face-to-face.	Owen: I remember my grandma teaching me to plant seedlings in little pots. I would have been, maybe, six years old? She said I was the best person to help her because I had small hands.
8 HPW	Same view, closer in.	Owen: She didn't only teach you about growing plants, though, did she? Amal: She taught me that I was important. I mattered.

NB: most pages are made up of three full page width panels (FPW), six half page width panels (HPW), or a combination of the two.

Here are the resulting pages:

Figure 4.2 Page 1 from *Conversation with a Purpose* by Helen Kara and Sophie Jackson

Figure 4.3 Page 2 from *Conversation with a Purpose* by Helen Kara and Sophie Jackson

As you can see, some of my initial ideas were changed by the artist, Sophie Jackson, who had a much better sense of what was needed visually. I was fine with that – a key point of the collaboration was for us to combine our ideas. Sophie sent me draft pages and occasionally I asked for small amendments to the visuals, or she asked for small amendments to the text. In the process, we figured it out together. The resulting short comic is free to download[50] and is being used in educational institutions around the world.

Comic making is very time-consuming, particularly the drawing parts. Professional comics makers often reckon to produce one finished page per day. At the time of writing, fees for comic artists vary from US$100 to US$325 per page, and for comic writers it's US$20 to US$125.[51] So hiring a professional artist or writer can be quite expensive. I was lucky to get to work with Sophie while she was still a student on the BA Cartoon and Comic Arts Degree at Staffordshire University, so her day rate was quite low. Staffordshire University is not far from my own home and workplace. If there is a course like that at a university near you, or you can work remotely with a student, then you may be able to arrange something similar.

Zines

Like comics, zines are useful for communicating complexity in accessible ways. Zines were originally handmade self-published booklets with tiny print runs of one hundred or fewer. Nobody is quite sure where they originated, perhaps because there are overlaps between zines and other early forms of self-publishing such as pamphlets and chapbooks. Now zines may be digital, known as ezines, or hard copy. They combine text and images in a creative, often collage-like style.

There are various kinds of zine such as fanzines and perzines. Fanzines are produced by fans of a public figure such as a sportsperson or a group such as a band, and handed out to other fans at events where that person or group is appearing. Perzine is short for 'personal zine' and is autobiographical.[52]

The publication rate of zines has increased enormously in recent decades,[53] perhaps at least in part because of the increase in digital options. Many libraries around the world now hold collections of zines for readers, artists, and others to use.[54] Some bookstores carry zines for sale. All of this could create the impression that zines are, or are becoming, mainstream. Yet this is in direct contravention of the whole point of zines. This is to offer opportunities for self-publishing, self-promotion, and communication with

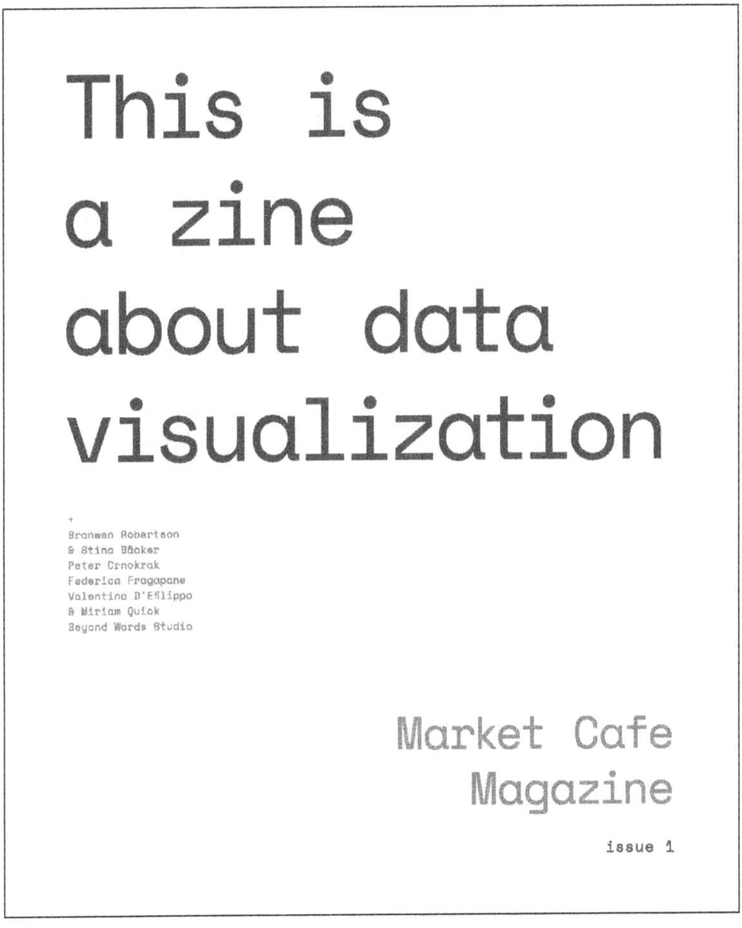

This is a zine about data visualization

Bronwen Robertson
& Stina Bäcker
Peter Crnokrak
Federica Fragapane
Valentina D'Efilippo
& Miriam Quick
Beyond Words Studio

Market Cafe
Magazine

issue 1

Figure 4.4 Cover of issue 1 of the *Market Cafe Magazine*
© Piero Zagami and Tiziana Alocci

others who share the same interests, that are accessible to individuals and groups whom mainstream publishing and broadcasting organisations customarily ignore.[55] Zine making can provide an opportunity for deep and critical thinking outside hierarchies of power.[56]

This could create the impression that zines are nothing to do with professional life. However, zines are now being used in a range of professional contexts. Let's look at three examples. The *Market Cafe Magazine* is a hard copy zine, founded in Europe in 2016 by designers Tiziana Alocci and Piero Zagami, with small print runs of hand-numbered issues which are sent out by post to purchasers. It is written mostly by designers for other designers and anyone else interested in ways of visualising data.[57]

The Sociological Fiction Zine, aka *So Fi Zine*, was founded in 2017. It is a digital zine, founded and run by Australian academic Ash Watson. The aim is to counteract 'the exclusive practices of academic publishing'.[58] Despite its name, the zine publishes poems and visual arts as well as short fiction, all inspired by social science. *So Fi Zine* is digital and free, with easy printing instructions for those who want a hard copy.

The digital PO Box zines are published by Project Orange, a small architecture and interior design practice in the UK founded by Christopher Ash and James Soane. PO Box 1 was published in 2010. It is a collection of essays which aims to bridge the gaps 'between theory and practice, between the world of ideas and the world of building'.[59] PO Box 1 was shortlisted for the Royal Institute of British Architects' President's Award for Research in 2011.[60] PO Box 2 was published in 2014 and examined representation through drawing; PO Box 3 in 2016 focused on housing; and PO Box 4 in 2020 considered the relationship between architecture and the climate crisis. Each zine had a different editor and a range of contributors, and all the zines can be downloaded from the Project Orange website.

One interesting aspect of these professional zines is that they are all multi-authored texts or edited collections, in contrast to the more conventional types of zine which are usually created by one person working alone. This serves to underline a key point about zines: they have hardly any rules. They can be long or short, monochrome or coloured, created by one person or more. They do usually include both text and images, and – like comics – often have pages in multiples of 4 (though this is not essential for digital zines which are not also designed to be printed). Apart from these last two points, anything goes.

Whether zines are made professionally or by amateurs, they are part of wider conversations just as much as newsletters or blogs or any other such format. In the workplace, a zine could be produced as a record of a conference or awayday, to support an induction programme, or in any other context where there is a need to store and/or share information.

How to make zines

There is no right or wrong way to make a zine. However, there are some guidelines. The first and most important thing is to figure out what you want to say, and then decide how best to put that message across in your zine.[61]

Overall, there are eight straightforward steps for making a zine:

1. Decide on a topic.
2. Generate text about that topic.
3. Find images to accompany your text.
4. Make an eye-catching cover.
5. Include your name and, if you wish, some contact information.
6. Finish creating your zine.
7. Upload, if it's an ezine, or print copies.
8. Distribute your zine.

You can do this individually or in collaboration. Making zines with other people can be 'a democratic and inclusive means of creatively ... exploring messages, experiences and ideas'.[62]

Barriers to working with comics and zines

Some people find cartoons, comics, and zines difficult to read because, as one person said to me, 'I don't know where my eyes should go'. Reading these formats is different from reading print where you start in one corner of the page and follow each line systematically until you reach the opposite corner. Some people need to learn to read graphic formats – sometimes known as 'comics literacy' – but it is not a difficult skill to acquire.[63] Novice comics readers often read all the text on a page first and then look at the pictures, as you might do with a conventional illustrated book. This is not an advisable approach, as the text and the pictures work together rather than the pictures supporting the text. Hannah Berry, an eminent comics artist and former UK Comics Laureate, says you need to absorb all the image and text in one panel before moving to the next. You can still read from one corner to the opposite corner, and you will usually read all the panels on one level first and then all the panels on the next level, and so on until you reach the end of the page.[64]

I grew up reading comics and have always enjoyed the freedom of finding my own track around and through, but I can understand that this does not appeal to everyone. Also, although digital comics are definitely a Thing, they do not work well on e-readers, which can be a barrier for people who prefer to read on those devices. However, they do work on most tablets and laptops. So there are some known barriers to working with comics, which probably also apply to some extent to cartoons and zines, and there are also ways of overcoming these barriers.

Conclusion

In this chapter I have demonstrated that cartoons, comics, and zines already play a key role in a range of professional settings. These formats also have a great deal of untapped potential for use in the world of work. If you would like to explore one or more of these formats for use in your own workplace, but feel unsure of your way, I recommend finding and reading some good examples. Any cited in this chapter are worth checking out, or you could look for comics online that have good reviews, or ask a librarian or bookseller for their recommendations.

A comic

Figure 4.5 *A Writer Writes* – comic by Helen Kara and Iqra Babar

A zine

Figure 4.6 Zine page 1 – cover page

Figure 4.7 Zine page 2

Figure 4.8 Zine page 3

Figure 4.9 Zine page 4

Figure 4.10 Zine page 5

Try it yourself

Make a four-page comic or zine about an aspect of your work or workplace. You can create your comic or zine by hand or digitally, or use a combination of both approaches. Your pages may be standard size, or you can fold a standard sized piece of paper in half to form a four-page comic or zine with half-size pages. Remember that the first page needs to be your cover. Use the guidance in this chapter to help you make your comic or zine. Remember drawing skills are not required: stick figures, or blobs with eyes, are perfectly acceptable. And don't forget to have fun!

Notes

1 Dorota Pawlik, 'Māori's ritual body embellishments', *Journal of Martial Arts Anthropology*, 11.4 (2011), pp. 6–11.
2 Stephen Zagala, 'Vanuatu sand drawing', *Museum International*, 56.1–2 (2004), pp. 32–35.
3 Khawaja Muhammad Saeed, 'Islamic art and its spiritual message', *International Journal of Humanities and Social Science*, 1.2 (2011), pp. 227–234.
4 Pawlik, 'Māori's ritual body embellishments'.
5 Randy Duncan, Matthew J. Smith, and Paul Levitz, *The Power of Comics: History, Form and Culture* (2nd edn) (Bloomsbury, 2015), p. xvi.
6 Randy Duncan, Michael Ray Taylor, and David Stoddard, *Creating Comics as Journalism, Memoir & Nonfiction* (Routledge, 2016), p. 1.
7 Scott McCloud, *Understanding Comics* (HarperCollins, 1993), pp. 10–15.
8 Harriet Earle, *Comics: An Introduction* (Routledge, 2021), p. 1.
9 Simon Grennan, 'Marie Duval: A Victorian cartoonist', in *The Inking Woman: 250 Years of Women Cartoon and Comic Artists in Britain*, ed. by Nicola Streeten and Cath Tate (Myriad Editions, 2018), p. 13.
10 McCloud, *Understanding Comics*, p. 3.
11 Seb Emina, 'In France, comic books are serious business', *New York Times*, 29 January 2019, www.nytimes.com/2019/01/29/books/france-comic-books-angouleme.html [accessed 17 August 2023].
12 Duncan, Smith, and Levitz, *The Power of Comics*, p. xi.
13 Earle, *Comics*, pp. 4–5.
14 Scott McCloud, *Reinventing Comics* (HarperCollins, 2000), p. 28.

15 Marietjie Botes, 'Using comics to communicate legal contract cancellation', *The Comics Grid: Journal of Comics Scholarship*, 7.1 (2017), p. 14, https://doi.org/10.16995/cg.100
16 Duncan, Taylor, and Stoddard, *Creating Comics*, p. 44.
17 Greg Sadowski, *Action! Mystery! Thrills! Great Comic Book Covers 1936–1945* (Fantagraphics Books, 2012).
18 Some excerpts from *Army Motors* can be seen in the University of Nebraska online collections: https://mediacommons.unl.edu/luna/servlet/detail/UNL~113~113~748~1455575:Army-Motors,-excerpts--1942- [accessed 17 August 2023].
19 Denis Kitchen, 'Editor's note', in *Comics and Sequential Art: Principles and Practices from the Legendary Cartoonist*, ed. by Will Eisner (Norton, 2008), p. 157.
20 Helen Kara and Jenni Brooks, 'The potential role of comics in teaching qualitative research methods', *The Qualitative Report*, 25.7 (2020), pp. 1754–1765 (p. 1755), https://nsuworks.nova.edu/tqr/vol25/iss7/2 [accessed 18 August 2023].
21 Gemma Sou and John Cei Douglas, *After Maria: Everyday Recovery from Disaster* (University of Manchester/HCRI, 2019), https://hummedia.manchester.ac.uk/institutes/hcri/after-maria/after-maria-eng-web.pdf [accessed 7 December 2024].
22 'WCIP comics', Wellcome Centre for Integrative Parasitology, www.gla.ac.uk/research/az/wcip/engage/publicengagement/wcipcomics/ [accessed 18 August 2023].
23 Duncan, Taylor, and Stoddard, *Creating Comics*, pp. 18–26.
24 Sadiya Nasira Al Faruque, 'The unique power of comics: a comprehensive exploration of the visual-verbal medium and its impact on storytelling, communication, and culture', *International Journal of English and Studies*, 5.10 (2023), pp. 1–11.
25 Ian Williams, 'What is "graphic medicine"?', Graphic Medicine, www.graphicmedicine.org/why-graphic-medicine/ [accessed 18 August 2023].
26 'Graphic Medicine International Collective', Graphic Medicine, www.graphicmedicine.org/graphic-medicine-international-collective/ [accessed 18 August 2023].
27 Scott McCloud, 'The Google Chrome comic', ScottMcCloud.com, https://scottmccloud.com/googlechrome/ [accessed 18 August 2023].
28 Robert G Weiner and Carrye Kay Syma, 'Introduction', in *Graphic Novels and Comics in the Classroom: Essays on the Educational Power of Sequential Art*, ed. by Carrye Kay Syma and Robert G Weiner (McFarland, 2013), pp. 1–11.
29 Sou and Douglas, *After Maria*.

30 Gareth Brookes, *A Thousand Coloured Castles* (Myriad Editions, 2017).
31 Christina Blanch and Thalia Mulvihill, 'The attitudes of some students on the use of comics in higher education', in *Graphic Novels and Comics in the Classroom*, ed. by Syma and Weiner, pp. 35–47.
32 Paul A. Aleixo and Krystina Sumner, 'Memory for biopsychology material presented in comic book format', *Journal of Graphic Novels and Comics*, 8.1 (2017), pp. 79–88, https://doi.org/10.1080/21504857.2016.1219957
33 Scott McCloud, *Making Comics* (HarperCollins, 2006), p. 4.
34 Kara and Brooks, 'The potential role of comics', p. 1761.
35 Christopher Murray and Golnar Nabizadeh, 'Introduction: "Public information comics"', *Law and Humanities*, 17.1 (2023), pp. 6–16 (p. 8), https://doi.org/10.1080/17521483.2022.2150171
36 BBC, 'M&Ms replacing spokescandies with comedian Maya Rudolph', *BBC News*, 23 January 2023, www.bbc.co.uk/news/world-us-canada-64380510 [accessed 7 October 2023].
37 Meghan Parker, *Teaching Artfully* (Clover Press, 2021).
38 Apostolos Doxiadis, Christos H Papadimitriou, Alecos Papadatos, and Annie Di Donna, *Logicomix: An Epic Search for Truth* (Bloomsbury, 2009).
39 Joe Sacco, *Palestine* (Jonathan Cape, 2003).
40 Patti LaBoucane-Benson and Kelly Mellings, *The Outside Circle* (House of Anansi, 2015).
41 Mirion Malle, *This is How I Disappear*, trans. by Aleshia Jensen and Bronwyn Haslam (Drawn & Quarterly, 2022).
42 Murray and Nabizadeh, 'Public information comics', p. 6.
43 Stuart Medley, Christopher Kueh, and Bruce Mutard, 'Letting the picture tell the story: using comics capture content as a research method', in *Handbook of Creative Research Methods*, ed. by Helen Kara (Bloomsbury Academic, 2024), pp. 163–174.
44 Nur Akcanca, 'An alternative teaching tool in science education: Educational comics', *International Online Journal of Education and Teaching*, 7.4 (2020), pp. 1550–1570 (p. 1558), https://iojet.org/index.php/IOJET/article/view/1063 [accessed 7 December 2024].
45 Weiner and Syma, 'Introduction', p. 1.
46 Sandrina Ritzmann, Annette Kluge, and Vera Hagemann, 'Using comics as a transfer support tool for crew resource management training', in *Proceedings of the Human Factors and Ergonomics Society Annual Meeting*, 55.1 (2011), pp. 2118–2122.
47 Kara and Brooks, 'The potential role of comics'.

48 Amber Nalu and James P Bliss, 'Comics as a cognitive training medium for expert decision making', in *Proceedings of the Human Factors and Ergonomics Society Annual Meeting*, 55.1 (2011), pp. 2123–2127.
49 Kara and Brooks, 'The potential role of comics', p. 1761.
50 Helen Kara, *Conversation with a Purpose*, Helen Kara, 12 June 2018, https://helenkara.com/2018/06/12/conversation-with-a-purpose/ [accessed 23 February 2024].
51 Creator Resource, 'Page rate finder – comic book publishers', *Creator Resource*, 13 June 2023, www.creatorresource.com/page-rate-finder-comic-book-publishers/ [accessed 30 November 2023].
52 Melanie Ramdarshan Bold, 'Why diverse zines matter: a case study of the people of color zines project', *Publishing Research Quarterly*, 33.3 (2017), pp. 215–228 (p. 219).
53 Joshua Barton and Patrick Olson, 'Cite first, ask questions later?', *Papers of the Bibliographical Society of America*, 113.2 (2019), pp. 205–216 (pp. 205–206).
54 Barnard College Zine Libraries, https://zines.barnard.edu/zine-libraries [accessed 1 September 2024].
55 Ramdarshan Bold, 'Why diverse zines matter', p. 216.
56 Tomás Boatwright, 'Flux Zine: Black Queer storytelling', *Equity & Excellence in Education*, 52.4 (2019), pp. 383–395, https://doi.org/10.1080/10665684.2019.1696254 [accessed 30 November 2023].
57 'The world's first magazine about data visualization', *Market Cafe Magazine* (undated), www.marketcafemag.com/about [accessed 23 August 2023].
58 'About', *So Fi Zine* (undated), https://sofizine.com/about/ [accessed 23 August 2023].
59 Project Orange publications, www.projectorange.com/publications [accessed 24 August 2023].
60 James Soane, 'Out of practice: theoretical speculations in and out of the business of architecture', *Architectural Design*, 89.3 (2019), pp. 48–53 (p. 51).
61 Clare Bonetree, 'Some tips', in *Doing It Together: An (Un)Guide to Making Zines with People*, ed. by Jean McEwan and others (2023), p. 14, https://changingrealities.org/zines/doing-it-together-an-unguide#open [accessed 7 December 2024].
62 Jean McEwan, (Unnumbered, untitled page inside front cover), in *Doing It Together*, ed. by McEwan and others (2023), p. 2, https://changingrealities.org/zines/doing-it-together-an-unguide#open [accessed 7 December 2024].

63 Samantha Golding and Diarmuid Verrier, 'Teaching people to read comics: the impact of a visual literacy intervention on comprehension of educational comics', *Journal of Graphic Novels and Comics*, 12.5 (2021), pp. 824–836, https://doi.org/10.1080/21504857.2020.1786419

64 Hannah Berry, 'How to read comics: a beginner's guide', BookTrust.org, 8 July 2012, www.booktrust.org.uk/news-and-features/features/2012/how-to-read-comics-a-beginners-guide/ [accessed 23 February 2024].

5

Dramatic writing

Introduction

Dramatic writing refers to play scripts, screenplays, and comedy writing. Play scripts and screenplays are largely based on talk, usually dialogue, though they may also contain monologue. The word 'monologue' is derived from the Greek for 'speaking alone', while 'dialogue' is derived from Greek words meaning 'speaks across, speaks between'. As this suggests, play scripts and screenplays are particularly useful for portraying people's inner worlds, interactions, and relationships.

Conventionally, a play script is written for a theatrical performance, typically on stage in front of an audience though plays are also delivered via radio, TV, and other media. The stage imposes constraints. Everything has to happen in one place, unless it is a high-budget production with scope for elaborate scene changes or a revolving stage. There are usually only a handful of actors, and it is difficult to make actions such as flying or swimming believable on stage.

Screenplays are usually written for films or TV shows. Here there are fewer constraints, particularly since the advent of computer-generated imagery. In a film, you can fly on a broomstick, live in a spaceship, or have a troupe of dancing ants in striped legwarmers. And when films are animated, anything goes. However, don't make the mistake of thinking fewer constraints mean you can be more creative. Although it may seem counter-intuitive, constraints can actually promote creativity (see Chapter 3 for more on this).

Although this book is about writing, it seems important to acknowledge that drama can also be improvised. Improvisation, or 'improv', is a theatrical technique which can be very entertaining on stage or screen. In the workplace, this is more commonly known as 'role play' which is often dreaded and derided, perhaps because it can be more difficult for amateurs to perform effectively than it is for professional actors or comedians. Even so, role play can be a very useful technique in helping people to prepare for unfamiliar or difficult interactions. And writing can also be improvised, for example, through the technique of freewriting (see Chapter 7 for details).

Improvisation is also key to comedy writing. Whether you see yourself as a funny person or not, you can learn to write comedy. Even if you are naturally funny, writing comedy takes time and effort. But comedy writing can be learned as we will see later in this chapter. First, though, we will consider the role of play scripts and screenplays in the workplace.

Play scripts and screenplays in the workplace

In the workplace, play scripts and screenplays may or may not be intended for production. A 'production' may be a full-scale play or film production, or a single monologue or recitation or short video, or anything in between.

Here is an example of the in-between. Cate Watson, an academic from the UK, studied home–school partnerships.[1] She became particularly interested in the experience of a mother whose child had been diagnosed with attention deficit hyperactivity disorder (ADHD), and in the different ways the mother and the professionals told their stories of this event. The professionals told the story of a deviant and dysfunctional family. The mother told the story of a series of unconnected events, beginning with a day when the child forgot their school tie. Inspired by the satirical work of eighteenth-century English artist William Hogarth, Watson wrote a series of satirical scenes between various characters to highlight the absurdities in the situation. She took her scenes to a conference to present her work, and asked people from the audience to volunteer to read the various parts, with no rehearsal or direction. This was well received, with audience members describing it as stimulating and 'powerful'.[2]

Another performance piece was created by Tara Goldstein and Jocelyn Wickett in Canada. Tara Goldstein was at the time a teacher educator and playwright; Jocelyn Wickett was an MA student and theatre artist. In their city there was a school shooting: a fifteen-year-old pupil, Jordan Manners, was fatally shot on school premises by other pupils. A thorough investigation was held and a comprehensive report produced which ran to four volumes encompassing almost six hundred pages with findings and recommendations. Goldstein and Wickett adapted this into a thirty-minute performance. First they had to choose which story to tell. There were a number of stories in the report, and they chose the story of 'zero tolerance', which is a particular approach to school safety – though they chose it because the approach had 'not made schools any safer or healthier'.[3] They incorporated the story of the last minutes of Jordan Manners' life, to show the reality of this tragic event alongside the discussion of policy. Some characters, such as teachers, parents, and pupils, were represented by small groups of people; others, such as a reporter and a school principal, were represented by individuals. Goldstein and Wickett held three readings of the script, with different audiences, who gave feedback which Goldstein incorporated as she revised her draft script after each reading. The fourth draft of the script was performed at the district's annual Safe Schools Conference, attended by five hundred teachers in training. The staging was simple but effective, designed to reflect the power dynamics among the different individuals and groups, with the investigators and school principal placed furthest from the audience while parents and pupils were closest to the audience. Again, audience members described the performance as 'powerful'.[4]

Another type of play script is for 'verbatim theatre', a form of documentary theatre where the dialogue is based on, or only uses, the words of real people. Watson did this to some extent: the dialogue in her scripts was based on the recordings of her research encounters, with changes made to ensure participants' anonymity. True verbatim theatre only uses real people's words, and those words are explicitly gathered for the purpose. Verbatim theatre was developed in Stoke-on-Trent, England, in the 1970s, by Peter Cheeseman, founding director of the city's Victoria Theatre.[5] One of the first verbatim plays he produced was *Fight for Shelton Bar*, about the potential impact of the proposed closure of a big steelworks in the city which had employed up to ten thousand people.[6]

Dramatic writing 115

This play was constructed from local people's recorded speech, and was performed at the Victoria Theatre for local people including steelworkers and then broadcast on BBC television.

It is perhaps no coincidence that these examples, like many in this book, come from education and arts workplaces where this kind of creativity is often viewed as more acceptable than it may be in some other workplaces. Yet play scripts and screenplays have a lot of potential in the workplace. They may be written, to provide vignettes, case studies, and illustrations of interactions, or performed, either in person or through video, to share results and achievements in conference and other presentations.

Why is play-script and screenplay writing useful in the workplace?

You may be wondering how on earth this can really be relevant to the workplace beyond education and academia. If so, it may surprise you to learn that scripts are used in many workplaces for such common tasks as sales, marketing, and customer support. (We saw an example in the introductory chapter from emergency medical dispatchers in the US.) These may be formal scripts, devised by managers to help employees become effective in their jobs quickly and easily, or to ensure consistency of service provision. Or they may be informal scripts, developed by people in the course of their work as they discover which forms of wording help with different situations. Formal scripts can be a very useful way to manage institutional knowledge which, otherwise, can often be lost as people move on from the organisation. But it is essential that formal scripts are regularly revised so they don't become outdated or stale, and this is where an understanding of play-script and screenplay writing can be very useful.

There are many other ways in which play-script and screenplay writing techniques can be useful in the workplace. Some examples are:
- An illustrative case study for a report, which could be written:
  o as a monologue to display an individual's experience;
  o in dialogue to convey interactions and relationships.

- A screenplay for a video, which could be for:
    - communication within your organisation;
    - sales;
    - training/professional development;
    - promotion.
- Scripted scenes for training purposes, to show potential interactions between people such as:
    - staff from different departments;
    - salesperson and customer;
    - teacher and pupil;
    - medical professional and patient.
- A script for a dramatised presentation, which could be:
    - to colleagues;
    - to the board;
    - to a client;
    - for a conference.

How to write a play script or screenplay

First you need to answer two key questions:

1. Play script or screenplay?
2. What is the story you want to tell?

The question of whether to write a play script or a screenplay may be answered by its purpose. If you are planning a dramatised presentation, you will need to write a play script; if you are preparing to create a video, you will need to write a screenplay. However, if your writing is not intended for performance or broadcast, such as if you are writing a case study for a report, then you can decide whether to use the play-script or the screenplay format.

Plays are divided into acts, often three or five acts, though they may have more or fewer – some plays have just one. Then each act is divided into scenes. The classic three-act play structure looks like this:[7]

Act 1: introduction
Act 2: conflict
Act 3: resolution

Within each act, there could be three scenes, like this:

Act 1: introduction

> Scene 1: exposition – introducing the main characters, their wishes, and the obstacles they face
> Scene 2: inciting incident – a call to action which challenges the main character
> Scene 3: decision – the main character's response to the call to action

Act 2: conflict

> Scene 4: initial consequences – the main character faces new obstacles to overcome
> Scene 5: further consequences – in working to overcome the obstacles they face, the main character causes something to go badly wrong
> Scene 6: reflection – how the main character decides to progress from here

Act 3: resolution

> Scene 7: despair – the main character ends up in a situation so bad that it is questionable whether they will escape and recover
> Scene 8: climax – either the main character somehow manages to escape from the situation, or they realise they are permanently trapped
> Scene 9: denouement – loose ends are tied up and a satisfying ending is created

This may seem complex but it reflects a very common, possibly universal, human approach to story creation and storytelling. It has been argued that we are all to some extent playwrights, even if the plays many of us construct are solely in the privacy of our own imaginations.[8]

The classic three-act structure set out above underpins many plays, films, and novels. It works well, but there is nothing sacred about this structure. There is no reason why an act should have precisely three scenes, any more than a play should have precisely three acts. Also, the way I have set it out above may give the impression that all acts and scenes are, or could or should be, of equal length. This is not the case. Each of the first and third acts often occupy around 25 per cent of the length, with the second act taking up the other 50 per cent – though this can vary too.[9] The initial and further consequences scenes may be very long; the climactic scene is often quite short.[10] Also, the terminology is not fixed: 'initial consequences' is also known as 'rising action', 'further consequences' is also known as 'midpoint', and so on.

Knowledge of this structure can be helpful for writers – and readers, and viewers – in understanding more about story construction. But it is unlikely that you will need to write a full-length play in your workplace. It is much more likely that you will find a reason to write a scene or two.

A scene is a 'unit of drama' which should 'act like a mini-play' with 'a rising action building to a climactic point and a falling back towards resolution'.[11] A scene can be very short. Here is an example from *Our Country's Good*, a play written in the late twentieth century by British playwright Timberlake Wertenbaker, based on a novel by Australian author Thomas Keneally called *The Playmaker* about the colonisation of Australia by the British. One of the characters in the play is a nameless Aboriginal Australian who comments throughout, offering a very different perspective from that of the British settler characters. Here is the whole of Scene 2:

> Scene 2: A lone Aboriginal Australian describes the arrival of the first convict fleet in Botany Bay on January 20, 1788.
> The Aborigine: A giant canoe drifts on to the sea, clouds billowing from upright oars. This is a dream which has lost its way. Best to leave it alone.[12]

This scene is a perfect example of a tiny mini-play. The arrival of the boat is the initial action; the beautifully imagined reaction of the Aborigine is the climax; the concluding warning serves as a resolution.[13]

The scenes you write are likely to be longer than this one but they should follow the same general structure. It may be useful to consider American screenwriting expert Robert McKee's definition of a scene:

> A scene is an action through conflict in more or less continuous time and space that turns the value-charged condition of a character's life on at least one value with a degree of perceptible significance.[14]

This tells us that a scene needs to: occur in a single place and time; include action involving conflict of some kind; and demonstrate a change for the central character. We can see this in practice in Wertenbaker's scene above. The scene is set at a beach on the Australian coast. The conflict is between European and Australian Aboriginal ideas and ways of life – this is quite subtle in the scene itself, but because the play is historical and we know about the appalling abuses of colonisation, we can see the conflict more clearly. The change for the scene's central character, the nameless 'Aborigine', is that now they know something they didn't know before. The scene also generates an emotional impact in two main ways. First, from the novel perspective of the Aboriginal character ('clouds billowing from upright oars' – it takes a moment to realise that this is a description of sails and masts). Second, from the character's conclusion that it is 'best to leave it alone', because we know that won't be an option and there are horrors ahead for this eloquent and dignified person.

When you are ready to write your scene, start with a list of the characters (aka the 'dramatis personae'), with a very brief description of each one. Then briefly describe the setting and the time. Here is an example from *Blackmail* by American playwright Lynn Snyder, written in the early twenty-first century.[15] The play is about a congressman who is being investigated because a Washington intern, with whom he was having an affair, has disappeared. This is the start of the script.

Characters:

Bruce Whitson – 54, United States Congressman from Ohio.
Gordon Whitson – 28, his son, candidate for Ohio Assembly.
Lisa Clement – 32, sister of a missing Washington intern.
Megan Whitson – 54, Bruce's wife.
Masked Man – non-speaking role.

Scene 1:

(Late afternoon. Akron headquarters of Ohio Congressman Bruce Whitson. Bruce and Gordon are on stage.)

Bruce: I went to the airport right from the police station and got the first flight out, so I could …

Gordon: The police station!
Bruce: So I could talk to you and your mom before …
Gordon: The police? Again? This is the second time they …
Bruce: Just let me …
Gordon: In a police car?
Bruce: What!
Gordon: Handcuffed!
Bruce: Handcuffed?
Gordon: When they take someone in for questioning …
Bruce: They didn't *take me* in for questioning. They asked me to come in.

In just seventy-one words of quick-fire dialogue, Snyder shows us that Bruce is in trouble with the police and that Bruce and Gordon have a prickly relationship. We also learn that Gordon has a mother, and that he can jump to conclusions. This demonstrates how play writing can convey a lot of information in a few words.

With screenplays, the storytelling is done more through visual elements than through dialogue. The beginning of the screenplay for the comedy cop buddy movie *Hot Fuzz*, written by Edgar Wright and Simon Pegg,[16] exemplifies this. V.O. stands for voice over, i.e., narration.

1 INT. METROPOLITAN POLICE STATION – FRONT DESK – DAY

POLICE CONSTABLE NICHOLAS ANGEL bursts through the entrance of a city police station and flashes his warrant card.

MALE VOICE (V.O.)

Police Constable Nicholas Angel.

2 INT. METROPOLITAN POLICE STATION – DAY

ANGEL strides down a corridor. His collar number reads 777.

MALE VOICE (V.O.)

Born and schooled in London.
Graduated from Canterbury

University in 1993 with a double
degree in politics and sociology.

INSERT: ANGEL at training college standing amongst dopey looking trainees. They wear navy tee shirts and shorts.

> MALE VOICE (V.O.)
> Attended police training college,
> displaying an impressive attitude
> in both field training and
> theoretical studies.

INSERT: ANGEL running in riot gear down an alley, dodging petrol bombs, storming a fake hostage situation, finishing an exam and holding the paper aloft.

> MALE VOICE (V.O.)
> Excelled way beyond peers, passed
> into the Metropolitan Police
> Service –

INSERT: ANGEL surrounded by the same dopey faces as before, this time in full uniform, at a graduation parade.

> MALE VOICE (V.O.)
> – and soon proved worth as an
> officer. Establishing both a
> popularity and an effectiveness in
> the community –

INSERT: ANGEL talking with elderly people, a Chinese family in their native tongue, young offenders in a hall.

> MALE VOICE (V.O.)
> – furthering his skills with
> elective training courses in
> advanced driving –

INSERT: ANGEL doing an elaborate skid in a police car.

> MALE VOICE (V.O.)
> – as well as pioneering the use of
> the mountain bicycle –

INSERT: ANGEL doing an elaborate skid on a police bike.

                    MALE VOICE (V.O.)

        – and raising officer's morale with
        an inventive use of desktop
        publishing –

INSERT: ANGEL pinning up various notices in bright colors; they read 'BIKE SHED', 'CANTEEN', 'HATE CRIMES'.

                    MALE VOICE (V.O.)

        – In 2001 began operations in a
        North London armed response unit,
        Whiskey, Bravo 7 –

INSERT: ANGEL bursts into a stairwell of an apartment block as part of a heavily armed response team.

                    MALE VOICE (V.O.)

        – and received a bravery award for
        efforts in the resolution of
        Operation Crackdown –

INSERT: ANGEL storms a room where a wild eyed CRACKHEAD holds a family hostage with a KALASHNIKOV. ANGEL responds fast, firing a short burst. His expression is one of shock.

                    MALE VOICE (V.O.)

        In the last twelve months alone,
        Has received nine special
        commendations, achieved the highest
        arrest record for any officer in
        the borough and sustained three
        injuries in the line of duty, most
        recently in December when wounded
        by a man dressed as Father
        Christmas.

INSERT: we see dashes of framed commendations, multiple cuffing and a violent altercation with a wild eyed St Nick.

You can see that the narration is economically written and serves to support the fast-moving visuals. The scene above lasts around one and a half minutes on screen, and builds a comedic, parodic picture of an outstanding young policeman in the UK's capital city.

The play-script and screenplay openings above demonstrate one of the similarities between the two formats: they are both character-led. Other people and their interactions and relationships generate compelling drama. Audiences will connect with characters who lack something they need, face tough choices, or struggle with difficult relationships. If you have a message to communicate through one of these formats, you need to convey that message through people's experiences, reactions, and choices. Beware of the 'info dump' where dialogue is used to inform the audience directly, rather than through characters and their actions and interactions. Here is an example:

Scene 1:
(Early morning, in the fancy office of company director Bill who is on stage with sales manager Andrea.)
Bill: Our sales figures for last year were good, right?
Andrea: Yes. Our reps did 10 per cent better than the year before. Our stores did 15 per cent better, and our door-to-door people did the same. Our overall net profit was fifteen million dollars.

This is tedious writing. It tells us almost nothing about the characters, except perhaps that they have a reasonably good professional relationship – and that is not very interesting. We do learn that last year's sales figures were good, and that the company is sizeable, multi-faceted, and profitable. But we learn this through a quite unrealistic piece of dialogue – the 'info dump'. A company director should be well aware of their company's sales figures, particularly for the previous year. And there is no drama, no emotion.

Here is a different version of the same scene.

Scene 1:
(Early morning, in the fancy office of company director Bill who is on stage with sales manager Andrea.)
Bill: Why didn't our reps do as well as our stores last year?
Andrea: I –
Bill [pacing]: They need to be selling more. Much more! [gestures wildly]
Andrea: But –
Bill: Our profits last year were only fifteen million dollars. It is Not. Good. Enough!

This has almost the same number of words as the previous version, and is much more compelling. We learn that the company director is agitated, perhaps a little unhinged, maybe greedy or in trouble. And we are beginning to see a professional relationship which appears rather unequal. Power imbalances are often interesting for audiences.

Screenplay or play script?

Figuring out which format is best is not always easy, even for professional writers whose work is performed on stage and screen.[17] So when you simply want to bring more creativity into your workplace writing it can be even more difficult. Here are some pointers to help you decide.

If you want to tell your story through dialogue, you probably need to use the play-script format. If you want to depict actions, then a screenplay format will be more effective. If you have a balance between the two then it depends on the type of actions. Interpersonal actions, such as slaps and kisses, are easily managed on stage, while more complex actions, such as parties and battles, are more suited to the screenplay format.

Another factor to consider is that plays, in common with many live performances, are personal and unique experiences. This applies even with plays which are performed several times a week: each performance is different, every audience is different. They may be similar but they are not the same. Screenplays, though, form the basis of films and videos which offer more distance and consistency. Screenplay-based products have the potential to reach huge audiences, though there are no guarantees. Do you want to convey a sense of intimacy and immediacy, or a feeling of something more considered and deliberate? If the former, a play script is probably more suitable; if the latter, screenplay format is likely to be more appropriate.

Your writing preferences and skills may also influence your decision. If you like writing dialogue, and you are good at it, you will probably be drawn to the play-script format. If you are good at thinking visually then screenplay writing may be more appealing,

Dramatic writing 125

with its scope for suggesting camera angles, types of shot (close-up, panning, etc.), and so on.

Perhaps the most important thing to consider is what you want to achieve. If your end goal is an in-person performance, such as a presentation to a board or a conference, then a play script will be most useful. If you are making a video then a screenplay is likely to be more help. If your aim is to produce a piece for inclusion in a document, either format is fine.

Comedy

There is a prevailing myth about comedy: only naturally funny people can make good jokes. This is a myth for two reasons. First, pretty much everyone is naturally funny at times.[18] Second, if you can learn to write, you can learn to write comedy.[19]

That said, writing comedy isn't easy – but there are tips and tools to help you. One key tip is that you need to be prepared to generate a lot of ideas and then discard most of them. Quantity leads to quality. Anyone can learn to write jokes, but nobody can learn to write good jokes all the time.[20] Comedy writers write dozens, even hundreds of jokes for every one they use or sell.

Another tip is to avoid clichés. There are some humour-related clichés which are really worn out, such as 'See what I did there?', 'I'll get my coat' and 'If I tell you, I'll have to kill you.' Falling back on a cliché is often tempting, and sometimes it's worth using one or two for productivity's sake when you're generating ideas, because you can turn them into something more original later on by using your second thoughts (see Chapter 3 for more on these). But in general I would recommend avoiding clichés and making that avoidance a habit. (This applies to all writing, not just comedy – but it is perhaps most important here, because a worn-out cliché won't get laughs.)

I have read several books on writing comedy and they are all quite different. There are similarities along the lines of: hello, I am a successful comedy writer, and I'm going to tell you how I write comedy and therefore how you can write comedy. But the 'how to' bits vary a lot, which is interesting because it suggests that for comedy writing – as for writing in general – it's not just a case of 'what works' but 'what works for this writer at this time'. However, it does mean I can't synthesise and summarise the methods for you. Instead I'm going to recommend a text which was recommended to me by the British comedian and poet Kate Fox. It is *The Serious Guide to Joke Writing: How to Say Something Funny About Anything* by Sally Holloway, published by Bookshaker in 2010. Holloway offers a bunch of accessible joke-writing methods which interact with each other, and exercises to help you become familiar with those methods. She also provides a lot of useful insights into barriers to writing comedy and how to overcome them.

Another thing the books agree on is that comedy writing is hard work and can also be great fun. That is another point which could apply to writing more generally.

Conclusion

The play-script and screenplay formats have a lot to offer to workplace writing, as does comedy. At present these seem to be the least used options of all those included in this book. I wonder whether this is because they are not well understood by many people. As this chapter shows, they are not particularly complicated. Writing a whole three-act play or a screenplay for a full-length film would certainly be a challenge, and writing good jokes is difficult. But it is not beyond the reach of any halfway-competent writer to produce a scene in either format, or a decent joke, if that writer is willing to take the time and make the effort. For work-related writing purposes, one or two scenes or jokes will often be enough.

The start of a screenplay

> London bus stop, 1980s. Morning. Queue of people waiting for a bus. HELEN is in the queue, dressed in a university sweatshirt, ski pants and pixie boots, and wearing a backpack. She has burgundy lowlights in her neatly bobbed hair. Camera pans along the queue and turns to show a Routemaster bus pulling up to the stop. A couple of people jump off the platform at the back and several people from the queue get on.

Close-up on: HELEN walking through the downstairs of the bus, which is almost full, to an empty seat at the front on the left, in front of the window, next to ALISON, an attractive woman around Helen's age who is reading *The Color Purple*.

HELEN takes off her backpack and sits down as the bus starts to move, pulling her backpack onto her lap. She opens it and pulls out a notebook and pen, stows the backpack between her feet and turns to a blank page in her notebook. She puts the end of her pen in her mouth and sucks it absent-mindedly, thinking.

HELEN and ALISON from outside the window in front of them. ALISON is reading and HELEN is writing. They are both taking quick sideways looks at each other, though not at the same time.

HELEN writing and ALISON reading from inside the bus. There is a sound of someone vomiting. Helen winces and hunches into her seat.

ALISON

You OK?

HELEN

(weakly) I'm not so good with …

HELEN gags, covering her mouth with her left hand and gesturing behind her with her right hand.

ALISON

Hold your nose. And talk to me. It'll help.

HELEN holds her nose. The vomiting sounds stop. HELEN relaxes a little.

ALISON looks behind her.

ALISON

It's a young boy, with his mother. She has a bag, there's no mess. They're getting off.

The bus comes to a stop. HELEN is still holding her nose.

HELEN

(muffled) Thank you.

> ALISON
>
> You can go back to your essay now.

HELEN releases her nose and takes a cautious breath, then a deeper one.

> HELEN
>
> Aaahhh, lovely. Diesel and the huddled masses. And your gorgeous perfume, of course.
>
> ALISON
>
> Babe, Faberge.

HELEN's mouth drops open for a moment, then closes again.

> HELEN
>
> That's the perfume?

ALISON grins and gives Helen a flirty look.

> ALISON
>
> Did you think I meant you?
>
> HELEN
>
> (sheepish grin) Just for a moment.

HELEN goes back to her writing, and ALISON to her reading. They are both grinning to themselves.

Try it yourself

Choose one of these two options – or do them both if you like!

Option 1. This enjoyable technique can help you find new insights into your work in general, or a particular problem you are facing. Think of a fictional character you know well. This could be a character from a film or a play, a book or a comic, a soap opera or a rock opera, an advertisement or a fairy tale – whatever you like. Tell that character about your work, or your problem, in your own written words. Then write down their reply, and continue to create a written dialogue between the two of you until it feels complete. Then decide whether it would be a staged or filmed scene, and add any stage directions or directors' notes you wish. This is purely for your own use, so you can take the dialogue in any direction you like – or let it take you in any direction that may appear. You cannot do this wrong! It doesn't matter whether your dialogue is serious or playful, long or short, written in English or another language. The point is for you to unleash your creativity and have some fun with words and ideas. When you have written your dialogue, read it over and consider what you have learned from the process.

OR:

Option 2. Think of a memorable incident from your professional life and recreate it as a scene from a play or screenplay. First list your characters with a brief description of each one – just a few key points, for example, name, gender, race/ethnicity, approximate age, appearance. Give them pseudonyms and change some identifying details if you like. Then write the scene, including stage directions/directors' notes as appropriate.

Notes

1 Cate Watson, 'Staking a small claim for fictional narratives in social and educational research', *Qualitative Research*, 11.4 (2011), pp. 395–408.
2 *Ibid.*, p. 402.
3 Tara Goldstein and Jocelyn Wickett, 'Zero tolerance: a stage adaptation of an investigative report on school safety', *Qualitative Inquiry*, 15.10 (2009), pp. 1552–1568, http://doi.10.1177/1077800409343069
4 *Ibid.*, p. 1560.
5 Derek Paget, '"Verbatim theatre": oral history and documentary techniques', *New Theatre Quarterly*, 3.12 (1987), pp. 317–336.
6 Sarah Chapman, 'Buzzer signals last rites for men of steel', *The Sentinel*, 27 April 2000, www.thepotteries.org/shelton/newspaper.htm [accessed 16 December 2024].
7 Reedsy Editorial Team, ed. by Martin Cavannagh, 'The three-act structure: The king of story structures', *Reedsyblog*, 18 March 2021, https://blog.reedsy.com/guide/story-structure/three-act-structure/ [accessed 5 January 2024].

8 Joe Norris, 'Reflections on the techniques and tones of playbuilding by a director/actor/researcher/teacher', in *Handbook of Arts-Based Research*, ed. by Patricia Leavy (The Guilford Press, 2018), p. 293.
9 Robert McKee, *Story: Substance, Structure, Style and the Principles of Screenwriting* (Methuen, 1999), p. 219.
10 Reedsy, 'The three-act structure'.
11 Derek Neale, 'Staging stories', in *A Creative Writing Handbook: Developing Dramatic Technique, Individual Style and Voice*, ed. by Derek Neale (Black, 2009), pp. 73–92 (p. 86).
12 Timberlake Wertenbaker, 'From *Our Country's Good*', in *A Creative Writing Handbook*, ed. by Neale (Black, 2009), p. 338.
13 Neale, 'Staging stories', p. 87.
14 McKee, *Story*, p. 35.
15 Lynn Snyder, *Blackmail*, Lazy Bee Scripts (revised edn, 2013), www.lazybeescripts.co.uk/Scripts/script.aspx?iSS=1189 [accessed 5 January 2024].
16 Edgar Wright and Simon Pegg, *The Hot Fuzz*, https://app.studiobinder.com/company/580e85847e7982164664e844/collab/5fce92e9445bea11243db79a/projects/5fce80115add45129e167c1d/document/5fce801ee3bc4c11dd180378 [accessed 1 February 2024].
17 E. M. Welsh, 'Playwriting vs. screenwriting: What's the difference?', emwelsh.com, 16 August 2017, www.emwelsh.com/blog/playwriting-vs-screenwriting [accessed 16 February 2024].
18 Gene Perret, *The New Comedy Writing Step by Step* (Quill Driver Books, 2007), p. 6.
19 Scott Dikkers, *How to Write Funny: Your Serious, Step-by-Step Blueprint for Creating Incredibly, Irresistibly, Successfully Hilarious Writing* (self-published, 2015), p. 15.
20 *Ibid.*, p. 20.

6

Epistolary and digital writing

Introduction

'Epistolary' comes from the Latin word for letter (epistola) and the Greek word for message (epistole). If you are familiar with the Bible, you will be aware of the 'epistles' written by St Paul and others; in case you are not, they were letters to early Christian believers and new churches. These days 'epistolary' covers all forms of written communication from one person to another, such as letters, emails, and messaging. Digital writing has similar features and includes social media posts and comments, and writing for the web. Of course emails and messaging are also digital, so these categories are not separate but overlapping. They are linked by two main features: first, the communication is intended for a particular individual or group, and second, there is usually a physical distance between communicator and recipient(s).

Letter writing is often described as a 'lost art', but letters are the basis of all epistolary writing. Charities know that carefully crafted personalised letters can make a big impact which is why they are used for direct mail fund-raising campaigns. Letters also form the basis of other art forms, such as the stage show 'Letters Live', in which celebrities read out genuine letters from people past and present,[1] and the literary journal *The Letters Page* which publishes essays, stories, poetry, memoir, travelogue, and criticism, but only takes submissions in the form of handwritten letters.[2] The value we place on letters is probably because one person writing a letter to another is a reflective, intentional process, resulting in a unique artefact. The recipient of a personal letter knows the sender took some time and made some effort to create that letter for them.

This is why people value such letters highly, and those letters may become keepsakes. I have a 'nice letters' file where I keep kind letters and cards from clients and colleagues, so I can look at them when I'm feeling low to cheer myself up – and it works.

Writing is particularly important at a distance because your words are mostly how you will be judged. Yes, people can also judge the quality of your writing paper, or logo, or online ID – but the impression they form of you will be mostly created by your writing. So one key piece of advice is to write, read, think, and edit before sending. This is particularly important if you are having strong feelings about any aspect of the communication or its topic. It is also, paradoxically, important if you're working at speed. I once, absent-mindedly, signed off an email to a client with 'Love and kisses, Helen xxx'. When I realised my mistake, I was mortified and rang my client to apologise. Fortunately she saw the funny side.

Being creative with epistolary and digital writing can feel risky at times. Your common sense is likely to tell you if it's too risky. Taking small or medium-sized creative risks is often worthwhile, not least because doing so will strengthen your creativity.[3]

Letters

Letters can seem quite old-fashioned now, mostly used by organisations with creaking bureaucracies, but they do still have a role in our working lives. You may at times need to write a work-related letter of complaint, congratulation, or condolence. Applications for jobs, or funding, or in-depth courses often require covering letters, and a thank-you letter can be a nice touch at times. Congratulations, condolences, and thank-yous can also be written on cards, and, whichever format you use, handwriting is best for these, as they span the boundary between professional and personal communication. At least, handwriting is best if you are able to write by hand and your handwriting is reasonably legible; otherwise, stick to writing digitally.

Correspondence is essentially a form of conversation, but a slower, more thoughtful form than the verbal kind. Peter Elbow writes eloquently about this, pointing out that 'the leisure, privacy, and reflectiveness of writing'[4] can help you to figure out what you truly want to say and how you want to say it. He adds, 'The uninterrupted monologue of writing permits you to tell what it was really like, to say what you really felt, to finish the whole story, instead of so often being sidetracked by the give and take of conversation.'[5]

However, when you are writing letters in a work context, the tone should be professional, courteous, and clear, and the message should be as brief as possible.[6] Brevity is important because it shows that you value other people's time. However, again paradoxically, being brief may take you longer. There is a saying, used by various people through the centuries, along the lines of 'I'm sorry this letter is so long, I didn't have time to write a short one.'[7] The first draft of any letter can usually be edited down. Because of this, I write everything digitally to begin with, even if I plan to write it by hand in the end. This makes it easy to edit and shape my letter so I can be sure I am saying what I truly want to say.

Another element worth spending a little time on is accuracy. Check your spelling, punctuation, and grammar, and make sure there are no errors. If you are writing digitally, automatic spell checkers can help, but don't rely on them entirely because they will miss typos such as 'she had to disciple her dog' or 'hero and heroin' or 'he came form New England'. Make sure you have used the correct salutation for the person you are addressing, and that you have spelled their name right. For people you talk about in your letter, make sure you use their preferred pronouns, or – if in doubt – a neutral pronoun. The occasional typo is bound to slip through but your letters should be as error-free as possible. Lack of attention to detail does not look good in the workplace, and this is another area where people will be judging your professional competence.

These procedural points are important, but they risk obscuring the potential for creativity in letter writing. The key to this is to take the time to express what you want to say in your own words. Of course some business letters can be short and simple, even formulaic, and that's no problem. If a financial services provider writes a letter to inform their customers about a change to their terms and conditions, that letter doesn't need to be creatively written. Writing creatively is always an option, and some customers might appreciate a creatively written letter – it could even generate positive publicity for the organisation – but others might find it irritating, and no doubt some wouldn't notice or care.

Some workplaces use letters in purposeful ways. The charity Letters to a Pre-Scientist (LPS), founded in North Carolina, US, matches professional scientists with middle-school students in US low-income communities so they can exchange letters at set times during a school year. LPS' mission 'is to facilitate one-on-one connections to humanise STEM professionals, demystify STEM career pathways, and inspire all students to explore a future in STEM'.[8] In the early COVID-19 pandemic, LPS had to move to email for a while, but when they asked the schools they work with whether they would prefer to stick with email or go back to letters, all the schools wanted to go back to letters.[9] On pp. 135–137 is an example of a letter written to a 'pre-scientist' with annotations.

Emails

Emails have developed from letters so many of the same points apply. One difference is that emails are, by definition, always digital, and can be much briefer than letters. However, this depends on the context. I work with colleagues across Europe for the European Commission. Our email correspondence is always very polite, and set out like a letter, from 'Dear [name],' to 'Best wishes/kind regards, [name]'. I also work with UK publishers who, once you have got to know them, are happy with very brief emails, sometimes just a phrase such as 'Fine by me' or 'Thank you', or even a single word such as 'Yes' or 'OK'. Publishers deal with a huge number of emails so this level of brevity is helpful for them – and, I have to say, for me too.

This, again, requires some attention. If you are emailing someone you don't know, always use your most polite approach until you find out what their preferences are. When different correspondents use different approaches, take care not to mix them up, particularly if you are working in haste. This only requires 'a moment of thought'[10] but when you are under pressure it can be easy to forget to take that moment, and so potentially end up offending a correspondent, or at least not appearing as professional as you would wish.

Careless or impolite emailing can lead to misunderstandings and even conflict,[11] which is not desirable in a workplace – or anywhere

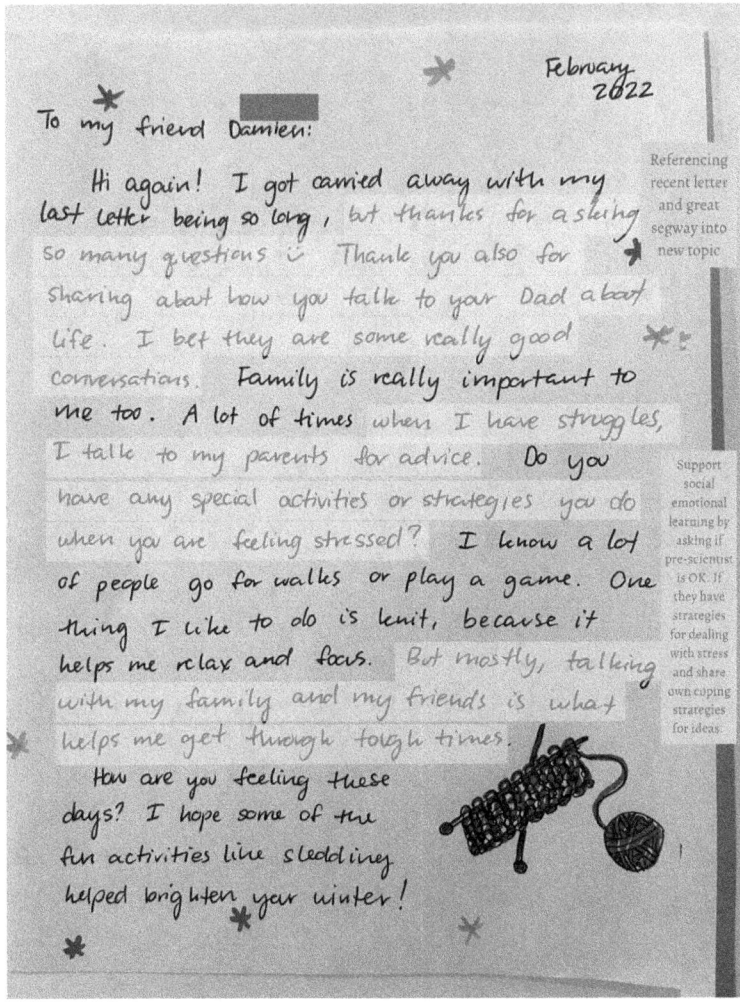

Figure 6.1 Letter to a pre-scientist page 1.
Reproduced by permission of Letters to a Pre-Scientist.

else. It is easy to misinterpret an email you receive because you only have the sender's words to go on. You can't draw information from people's facial expressions, body language, etc. as you would when interacting with them in person. So when you receive an email, remember that words alone can be quite ambiguous, and if necessary consider giving the sender the benefit of the doubt.

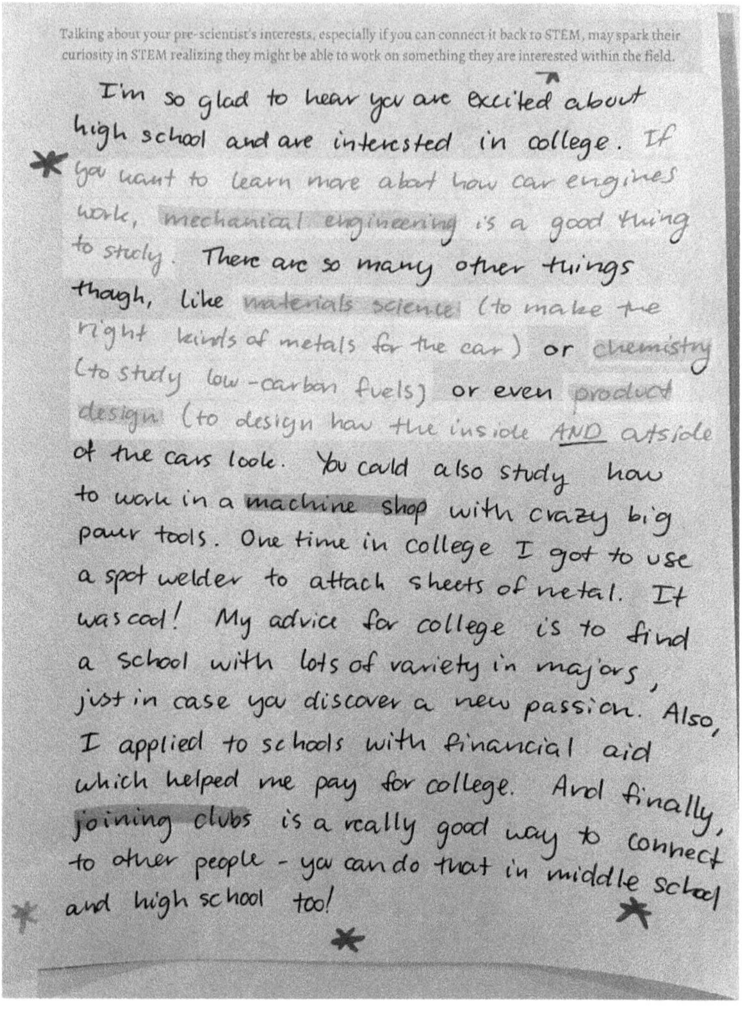

> Talking about your pre-scientist's interests, especially if you can connect it back to STEM, may spark their curiosity in STEM realizing they might be able to work on something they are interested within the field.

> I'm so glad to hear you are excited about high school and are interested in college. If you want to learn more about how car engines work, mechanical engineering is a good thing to study. There are so many other things though, like materials science (to make the right kinds of metals for the car) or chemistry (to study low-carbon fuels) or even product design (to design how the inside AND outside of the cars look. You could also study how to work in a machine shop with crazy big power tools. One time in college I got to use a spot welder to attach sheets of metal. It was cool! My advice for college is to find a school with lots of variety in majors, just in case you discover a new passion. Also, I applied to schools with financial aid which helped me pay for college. And finally, joining clubs is a really good way to connect to other people – you can do that in middle school and high school too!

Figure 6.2 Letter to a pre-scientist page 2.
Reproduced by permission of Letters to a Pre-Scientist.

Brevity is generally useful in work-related emails. Some people commit to keeping each email to five sentences or fewer,[12] though for others this is too much to ask. Either way, it is always worth taking a few moments to review an email you have written, before you send it, particularly with an eye to reducing its length. Look out for repetition, irrelevant content, and unnecessary information.

Epistolary and digital writing

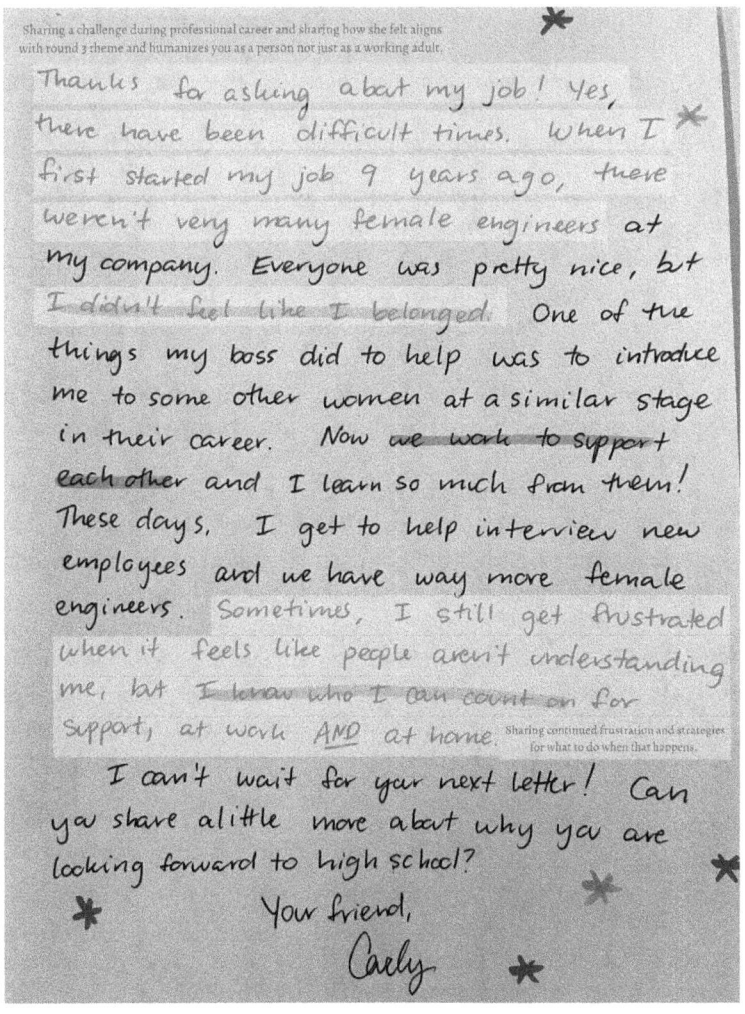

Figure 6.3 Letter to a pre-scientist page 3.
Reproduced by permission of Letters to a Pre-Scientist.

Even if you can just shave off a few words here and there, that is worth doing. The more you get into the habit of writing concisely the easier it becomes.[13]

The American businessman Elliot Bell offers useful advice.[14] He suggests re-reading an email before you send it, while considering these three questions:

1. Is the point you are making in this email clear and easy to understand?
2. Are there any words/phrases/sentences that don't add to the point you are making and so can be deleted?
3. Are there any words/phrases/sentences you can simplify?

Use the subject line strategically. If you are asking someone to do something, signal this in the subject line, and be as specific as possible. Say you want someone to complete an evaluation form. If you put 'evaluation' in the subject line, that's relevant but not specific enough. You could put 'evaluation – please complete' and that would be better. Better still would be 'three-question evaluation – please complete'.[15] That way you are giving the recipient a clear idea of what you want to communicate before they have even opened the email.

That said, don't fire off lots of small emails on the same topic, because that is tiring and confusing for recipients. If you're working through a process, gather the points you want to make and include them all in a single email, even if that makes it a long one.

Whatever its length, make sure your email is well structured and not a chaotic ramble. A well-structured email takes a little time to craft, but this is worth doing for two reasons. First, it makes life easier for the recipient(s), and second, it showcases your communication skills.[16] Poorly written emails generate more emails for clarification which is unnecessary.[17] Plus, taking the time to give your email a helpful structure will also enable you to make it more creative.

There is also scope for creativity in out-of-office messages. In December 2017, the British comedian and poet Kate Fox wrote an out-of-office message which read: 'May your joy and your emails go unchecked this holiday time.' This is ideal: short, humorous, and warm, so no doubt also memorable. In fact it was memorable enough for me to track it down much later and obtain Kate's permission to include it in this book. The poet Jonathan Davidson takes a different approach. Each weekend he sets up an out-of-office message which contains a poem by someone else (with the author's permission, if it's still in copyright). The message also offers the option to subscribe to a monthly email which contains all the poems he used in his out-of-office messages that month. As well as being a poet, Jonathan is the chief executive of Writing West Midlands. This approach is a gentle and effective way of promoting poetry which is part of his job.

Email sign-offs, too, offer scope for creativity. If you're bored with 'best wishes' or 'kind regards', you might want to consider something like 'congratulations on reading all the way to the end of this email' or 'it will soon be Friday' or 'why not agree with me to save time?'. Or, if you have something to promote, 'don't forget to [sign up for X/buy Y/attend Z]'. Even stating your pronouns can be done creatively. Writing coach Jo Van Every has used 'she/her is close enough', and I have also seen something like 'he/him/hey you'. There are lots of other examples online if you want more inspiration. Remember, though, that these kinds of messages may come across differently than you intend, so make sure whatever you use fits with your professional persona. If in doubt, test it out on a few trusted people before using it for real.

Social media

Social media has a myriad of purposes: good and bad, legal and illegal, time-wasting and time-saving, and so on. It is often presented negatively in the press, though there are occasional exceptions. The British journalist and broadcaster Amol Rajan wrote in 2024 of social media that 'Those of us in my trade should remember it can generate communities and pullulate with kindness and creativity rather than conspiracy and contempt.'[18]

Social media changes frequently. Platforms change their ways of operating, new platforms come and old ones go, and the way individuals use social media changes over time. Social media can be a great asset for work – some people make entire careers from social media – though it can also be a liability: people have been sacked from jobs and removed from education as a result of their posts on social media. As I write, news in the UK is reporting the sacking of police inspector Philip Grimwade for making discriminatory and abusive posts on social media over a six-year period.[19]

So use social media judiciously and with care. Useful work-related functions of social media include marketing, communication, recruitment, and learning. Some institutions have social media policies, and some ban social media; if you work for an employer, or study with a university or college, that has such a policy or ban, you will have to comply with their requirements. Conversely, in professions where good networks and up-to-date information are highly valued, such as medicine and journalism, people may use social media a lot in the course of their work.

The impressions people form of each other on social media may be accurate or inaccurate. Information on social media is often quite brief, and even longer forms of social media, such as blogging, are often very partial and carefully constructed. Research studies have shown that where information about someone posting on social media is limited, people fill in the gaps for themselves. They exaggerate their impression of the writer positively if the limited information available is positive, or negatively if it is negative.[20]

This suggests that when we write on social media, it behoves us to take a positive approach, because that makes it more likely we will be viewed positively by our peers and superiors. There are other good reasons for this, too. Social media has a potentially huge reach and also a long tail. You can delete a post you have made, but if someone has taken a screenshot that post may live on for ever whether you like it or not.[21] So it is worth being careful with the tone of any post, particularly as nuances such as humour and irony are often missed or misunderstood online.[22]

That said, the constraints of social media, including word limits and etiquette, can, paradoxically, enable creativity and foster its development.[23] (See Chapter 3 for more on constraints enabling creativity.) This creativity does not only include or apply to writing, but also other features of social media such as images, videos, polls, GIFs, and so on, which are often combined with writing in social media posts. Social media can also be very useful for generating creative inspiration[24] and, of course, for sharing creativity with others.

Writing for the internet

Writing social media updates and blog posts are forms of writing for the internet, and there are others too, such as writing copy for websites and forums. People will mostly read this content on screen; it is rare that anyone prints out online writing to read in hard copy. What is more, many readers will be using a mobile device with a small screen. For most people, reading on screen is more taxing and tiring than reading on paper, and it is easier to be distracted by an appealing video or link. So you need to write in a way that makes reading as easy as possible. This means using plain language with no jargon, short sentences and paragraphs, punchy prose, and plenty of headings. Plain language can be helpful for readers whose first language is not the language in which you are writing. Headings structure your content and enable readers to find the sections which interest them most.[25] Keep your writing as short and simple as possible, not to dumb it down but to make the reading easier.[26] Lists and bullet points can be useful too. There are several readability checkers online such as a free-to-use app called Readable[27] where you can paste in text to check how easy it is to read and get tips for improving its readability.

Try to make your writing as compelling as possible. Use sensory language where appropriate, to engage readers' memories and emotions. Put key information at the start, and try to structure it in a way that will encourage people to read on. There are various ways to do this such as by presenting the information as a question, a surprising fact, or a summary of opposing viewpoints. This is sometimes called a 'hook' because it is designed to hook the reader in to the text and keep them there, like a fish on a line. Vary the lengths and structures of your sentences, because this makes writing more engaging.[28] Though don't write any really long sentences because those can be confusing. Try for a range of one to twenty-five words with an average of fifteen to twenty words,[29] and have some sentences with punctuation and others without, apart from the full stop at the end.

Aim for a conversational to chatty tone: casual and informal but polite. Whenever possible use images to support your writing, and include 'alt text' which describes the content of an image for people with visual impairments (you can see examples in this book). If you need to use a technical or other term that won't be widely understood, embed a link to a definition elsewhere on the web.

Being polite online

Virginia Shea formulated a set of rules for civil digital communication in the mid-1990s and they still hold good today. Here are her ten core rules.[30]

1. Remember the human. You are using a machine to help you communicate, you are not communicating with a machine. Communicate in a way you would like to be communicated with yourself. And don't write anything you wouldn't say to someone in person.
2. Behave as well online as you would to someone's face. Ensure all your communication is both legal and ethical.
3. Behave appropriately for the digital environment. In a forum for football fans it may be fine to swear; in an online jobs marketplace swearing would be unwise. If you're not sure what is appropriate, take time to observe the environment before you join in.
4. Respect other people's time and energy. Keep messages short and to the point. Don't copy other people in unless they truly need to be part of, or be informed about, the conversation you are having.
5. Care for your online reputation. Ensure that the content of your messages is coherent, reasonable, and well written.
6. Share expert knowledge within your online communications. That may be your own expertise or the expertise of others.
7. Don't perpetuate problems online. If you see people arguing, or engaging in a pile-on, don't get involved. And don't forward insulting messages even if they are funny.
8. Respect other people's privacy. If someone sends you a private message, or has a private area online which you can access, don't forward their messages to others. [In 2022 Rebekah Vardy, wife of British footballer Jamie Vardy, was ordered by a court to pay millions of pounds in costs. This was because Rebekah Vardy had leaked private posts to the *Sun* newspaper from the private Instagram account of Coleen Rooney, wife of British footballer Wayne Rooney.]
9. Don't abuse your power. If you have more power than someone else, whether that is technical prowess or other forms of expertise, don't misuse it to take advantage of others.
10. Forgive the mistakes of others. We have all made mistakes online. Be kind.

The risks of creativity

Although it is important to be creative where possible, it is also important to be aware of the potential associated risks, and particularly so when your writing will appear in public on the internet. Because creativity involves finding and using your own individual perspectives, it can lead to more self-disclosure than may feel comfortable to you.[31] This can also mean that criticism of your creative outputs might feel more like a personal rejection than professional feedback, even if the criticism is entirely constructive. Sometimes online criticism is not constructive at all, so be prepared in case you find yourself receiving harsh responses to your work. Remember that online writing cannot be retracted unless you are 100 per cent sure nobody in the world took a screenshot. The risk of being judged is a risk some people are not willing to take, and that is fair enough.

Creativity is antithetical to control, and can lead to the psychological state of 'disinhibition' where people have reduced control over their behaviours and emotions.[32] There is a rebellious aspect to creativity which can lead to undesirable consequences in the workplace. Self-control is essential for compliance with the professional and social requirements of any workplace. Yet to be creative, we need to loosen the bonds of self-control to enable the experimentation and play that creation requires.

Of course there are also potential rewards from being creative, such as greater feelings of autonomy[33] and closer relationships with colleagues.[34] But keeping a balance between the self-control to do well at work, and the creativity to do well at work, is not always easy. I am highlighting these risks here, not to deter you from being creative, but to try to equip you with the information you need to assess and mitigate any potential risks in your own workplace context.

Conclusion

Epistolary and digital writing offers enormous scope for creativity in form, content, and purpose. The key to all types of epistolary and digital writing is communication. As with all types of writing, it is a good idea to identify your reader or readers and keep them in mind as you write.

Whatever form of epistolary or digital writing you are using – letter, tweet, blog post, email, Facebook update, web page, and so on – you can apply macro or micro creative techniques. Of course there is no need to be creative all the time; if you're simply congratulating someone online for an achievement they have shared, or responding to an email asking if you can meet on a given date, it makes sense to be straightforward and brief. But there is always scope for creativity, and you can take that opportunity whenever you choose.

A letter

My desk

2024

Dear Reader

Thank you for reading my book. It was a pleasure to write it for you. In fact right now, for me, it is a pleasure to write you a letter. The format is old-fashioned these days, but when I was a young woman, before the internet and mobile phones, letters were incredibly important. People wrote letters a lot at that time, and even more in earlier times, in countries where writing was valued and there was sufficient wealth and will to create a postal infrastructure. During Victoria's rule and through the world wars, in much of the UK there were several collections and deliveries of letters each day, and the 'night mail' trains carried post between cities. The British-American poet W. H. Auden expresses the importance of letters in his poem *Night Mail*, written in 1936. Here is part of that poem:

> Letters of thanks, letters from banks,
> Letters of joy from girl and boy,
> Receipted bills and invitations
> To inspect new stock or to visit relations,
> And applications for situations,
> And timid lovers' declarations ...

Auden goes on to mention letters enclosing holiday photos and condolence letters, among others. These days, timid lovers' declarations are more likely to be made via instant message, condolences posted on social media, and holiday photos shared in an app. Letters are much more often 'cold and official' documents from businesses.

I would like this letter, by contrast, to be warm and personal. I can't make it fully personal because I don't know you. I don't know where you might be as you read this, or why you have chosen my book, or what you would like to gain. But I can tell you I value you enormously. Writing is a rather solitary profession, which suits me because I'm usually happy to sit at my desk, in my office, in my garden, and play with words on a screen. I enjoy writing. But sometimes it gets lonely, and then I think of you, and that cheers me up.

My hope is that this book will help you in your career. I hope it will help in practical ways, to make your work-related writing easier and better quality. I hope it will help you personally, perhaps by increasing your confidence. And I hope it will help you to have more fun as you work and/or study. I have never understood why 'fun' and 'work' should be seen as opposites; why 'fun' is something for leisure time, or children. Of course not every task or interaction can be fun, but I think we could have a lot more fun, a lot more of the time, if we only allowed ourselves to do so.

I had fun writing this. I hope you had fun reading my letter. I wish you joy of your writing.

Helen

Try it yourself

Think of someone connected with your work who you have something to say to that you haven't said. This could be because you think it would be inappropriate, or you're afraid of upsetting them or jeopardising your job, or simply because you haven't found a suitable opportunity. Write them a letter to say what you want to say. Try to express your points as clearly as possible.

When you have written the letter, give yourself a little time to reflect on the experience.

- Did you learn anything?
- Do you feel different?
- Are you going to send the letter? Why?

Notes

1. 'Welcome to Letters Live', Letters Live, https://letterslive.com/ [accessed 14 December 2023].
2. 'Welcome to *The Letters Page*', *The Letters Page*, www.theletterspage.ac.uk/index.aspx [accessed 14 December 2023].
3. Ronald Beghetto, Maciej Karwowski, and Roni Reiter-Palmon, 'Intellectual risk taking: a moderating link between creative confidence and creative behavior?', *The Psychology of Aesthetics, Creativity, and the Arts*, 15.1 (2021), pp. 637–644 (p. 601), https://doi.org/10.1037/aca0000323
4. Peter Elbow, *Writing with Power: Techniques for Mastering the Writing Process* (2nd edn) (Oxford University Press, 1998), p. 96.
5. *Ibid.*, p. 97.
6. Sam Leith, *Write to the Point* (Profile Books, 2017), pp. 225–226.
7. Matt, 'If I had more time, I would have written a shorter letter', *Know Your Meme*, 31 October 2017, https://knowyourmeme.com/memes/if-i-had-more-time-i-would-have-written-a-shorter-letter [accessed 24 November 2023].
8. 'About us', *Letters to a Pre-Scientist* (undated), https://prescientist.org/about-us/ [accessed 19 January 2024]. STEM stands for 'science, technology, engineering, and mathematics'.
9. Omar Vera, 'You've got snail mail: how letters from STEM professionals are changing young lives', *Science Sandbox*, 15 December 2021, www.simonsfoundation.org/2021/12/15/youve-got-snail-mail-how-letters-from-stem-professionals-are-changing-young-lives/ [accessed 19 January 2024].
10. Leith, *Write to the Point*, p. 244.
11. Alan Sillars and Theodore E. Zorn, 'Hypernegative interpretation of negatively perceived email at work', *Management Communication Quarterly*, 35.2 (2021), pp. 171–200, https://doi.org/10.1177/0893318920979828
12. 'Five.sentenc.es', www.five.sentenc.es/ [accessed 20 March 2024].
13. Elliott Bell, 'Are your emails too long? (Hint: Probably)', *The Muse*, 19 June 2020, www.themuse.com/advice/are-your-emails-too-long-hint-probably [accessed 20 March 2024].
14. *Ibid.*
15. Jeff Su, 'How to write better emails at work', *Harvard Business Review*, 30 August 2021, https://hbr.org/2021/08/how-to-write-better-emails-at-work [accessed 19 December 2023].
16. *Ibid.*

17 Bell, 'Are your emails too long?'.
18 Amol Rajan, '"We need jungle" – Amol Rajan on how a University Challenge question spawned a remix craze', *BBC News*, 13 January 2024, www.bbc.co.uk/news/entertainment-arts-67955753 [accessed 19 January 2024].
19 Will Jefford, 'Nottinghamshire Police inspector sacked over abusive and targeted posts', *BBC News*, 20 November 2023, www.bbc.co.uk/news/uk-england-nottinghamshire-67479536 [accessed 24 November 2023].
20 Sillars and Zorn, 'Hypernegative interpretation'.
21 Leith, *Write to the Point*, p. 252.
22 *Ibid.*, p. 251.
23 Daniela Rezende Vilarinho-Pereira, Adrie A. Koehler, and Denise de Souza Fleith, 'Understanding the use of social media to foster student creativity: A systematic literature review', *Creativity: Theories-Research-Applications*, 8.1 (2021), pp. 124–147 (p. 141), https://doi.org/10.2478/ctra-2021-0009
24 Rafiq Elmansy, 'How social media affects your creativity', *Designorate*, 22 June 2015, www.designorate.com/how-social-media-affects-creativity/ [accessed 19 January 2024].
25 Donna Starks and Margaret J. Robertson, *50 Things to Think About When Writing a Thesis: Paving Your Own Path to Submission* (Routledge, 2024), p. 70.
26 Leith, *Write to the Point*, p. 247.
27 Readable app, https://app.readable.com/text/ [accessed 19 January 2024].
28 Starks and Robertson, *50 Things to Think About*, p. 76.
29 Plain English Campaign, *How to Write in Plain English* (Plain English Campaign, undated), www.plainenglish.co.uk/files/howto.pdf [accessed 23 February 2024].
30 Virginia Shea, 'The core rules of Netiquette' [excerpts from *Netiquette*], Albion.com (undated), www.albion.com/netiquette/corerules.html [accessed 13 November 2023].
31 Jack Goncalo and Joshua Katz, 'Your soul spills out: The creative act feels self-disclosing', *Personality and Social Psychology Bulletin*, 46.5 (2020), pp. 679–692, https://doi.org/10.1177/0146167219873480
32 Olga Khessina, Jack Goncalo, and Verena Krause, 'It's time to sober up: The direct costs, side effects and long-term consequences of creativity and innovation', *Research in Organizational Behavior*, 38 (2018), pp. 107–135 (p. 112), https://doi.org/10.1016/j.riob.2018.11.003

33 Sahoon Kim, Jack Goncalo, and Maria A Rodas, 'The cost of freedom: creative ideation boosts both feelings of autonomy and the fear of judgment', *Journal of Experimental Social Psychology*, 105 (2023), article 104432, https://doi.org/10.1016/j.jesp.2022.104432
34 Goncalo and Katz, 'Your soul spills out'.

7

The personal is professional

Introduction

We think of 'writing' as a task, as something onerous, perhaps rarefied; maybe a job for other people. Yet writing occupies more of our lives than we realise. Social media updates, emails, messages, letters, cards, forms to fill in – the list seems endless. Talking of lists, we also write those: to-do lists, shopping lists, bucket lists, packing lists, reading lists, and so on. These kinds of everyday writing are so common that we barely even notice them or think of them as writing.

As we saw in Chapter 1, writing is an emotional process, yet the emotional component of writing is rarely discussed.[1] Actually, everything we do has an emotional component, because we can't remove the parts of ourselves that generate and feel emotions. But writing seems to be a particularly emotional business. And the emotions evoked by writing in the workplace rarely seem to be positive. A psychologist specialising in emotions has described writing as hard, stressful, guilt-inducing, frustrating, complicated, and 'unfun'.[2] Yet some writers love writing.[3] Luckily for me, I'm one of them. If you're not, I hope this book will help you to enjoy writing at least a little more than you do now.

As well as playing an important role in our working lives, writing can also be highly personal. If you keep a diary, or write poems or stories or screenplays for your own amusement or for publication, you will know this yourself. Even so, you may not be aware that writing has three key roles to offer: it can be a teacher, a therapist, and a friend.[4]

Writing as teacher

People often assume that writers think first and then write down what they have thought. Sometimes this is indeed the sequence of events. But it is also true that, at other times, we learn from what we write – even from what we have only just written. A journalist friend of mine says 'I don't know what I think till I see what I write.' For me, the process is not so absolute, but I understand what she is saying. There have certainly been times when I have written something, then reacted to it with surprise and self-questioning: Is that what I really think? Is that what I mean? Is that what I want to say?

Freewriting can be helpful here.[5] This technique requires an active prompt such as 'I want to say ... ' or 'I care about this because ... ' and a pre-defined short period of time, say five minutes (though it could be ten minutes, or three, or seven). During this time you write freely and quickly in response to the prompt. You start by writing the prompt itself and then go on wherever that takes you, without changing or editing what you produce. If you get stuck, write out the prompt again, as many times as you need, until it takes you somewhere else. Keep your pen moving on the page or your hands moving on the keyboard. Don't worry about spelling, punctuation, or grammar.

Freewriting is a useful way to overcome our own internal barriers and critics.[6] If you're used to certain kinds of workplace writing with particular conventions, such as legal, academic, or technical writing, then freewriting may feel quite unnatural, even daunting, to begin with. Writing experts Sally O'Reilly and Linda Anderson recognise this, and also promise that freewriting can be productive and rewarding:

> It can feel uncomfortable, especially at first. You may feel that what you are writing is silly or unseemly or banal. You may feel a strong urge to stop or control it. But don't. You will often be surprised, even delighted, by the liveliness and power of the ideas and words that will emerge.[7]

I suspect freewriting is a way to access our 'precategorical realm'[8] or subconscious, albeit only briefly – but those brief flashes of insight can be extraordinarily illuminating. When you read back over your freewriting, much of it will be rambling and irrelevant, but here and there you will find a little nugget of inspiration. It can also help you to solve problems, such as – with a prompt like one of those suggested above – figuring out what you want to say, or why, in the section, chapter, report, or whatever it is you are writing.

At the other end of the spectrum, in long forms of writing, the writing process itself can teach us what we want to say and how. This happens as we work through the various drafts. It is almost impossible to produce a piece of writing that is both long and good quality without doing at least three drafts. Some writers refer to the first draft as the 'vomit draft': you get it all out, then you feel better, then you clean it up. Another common saying is that you write the first draft to tell yourself the story, and in the second draft you figure out how to tell that story to other people.

There is an analogy I like to use when I am teaching this process. Think of a potter making a vase. First she has to find her raw material: either dig some clay out of the ground, or buy it from a shop. Then it needs cleaning up (if dug out of the ground) and/or wedging to remove any small air bubbles that could cause problems during firing. When her raw material is ready, the potter can put it onto her wheel and shape it into a vase. Then, after firing, she needs to decorate the vase with glaze before it is fired again and finished. Writing a first draft is akin to finding and preparing your raw material so it is ready to work with. In the second draft, you shape your material, so people can see what you are producing even though it's not finished. And the decorating is akin to the polishing work which goes into a final draft. If you wanted to extend this analogy, you could refer to the firings as further drafts.

Sometimes inexperienced writers ask me why they can't get their writing right in one go. In theory this is possible, and in practice people do occasionally manage it, but it is quite rare. Peter Elbow says this is

> the dangerous method, because it puts more pressure on you and depends for its success on everything running smoothly. If you are out of practice or insecure or just a bit off your form, you can take longer trying to get something right the first time than you would have needed for writing roughly and then revising.[9]

This is because you are not allowing the writing to teach you in the process. As Peter Elbow points out, the need to get it right first time 'prevents the ingredients in your head from cooking, developing, progressing'.[10] Being determined to do all the learning in your own head first, and then write it down, might feel like a good idea – but more often than not you will find you cannot succeed.

The idea that writing can be a way to learn may seem counter-intuitive. After all, a teacher probably taught you to write in the first place, which instils a strong message that learning comes before writing. But I hope this section has at least begun to convince you that writing has an active role to play in learning. If you are still unsure, don't worry; you will find out for yourself as you write.

I should say that this applies not only to prose but to all writing. Glyn Maxwell, in his book *On Poetry*, explains how, when we embrace writing as a teacher, we end up:

> with words you didn't expect, echoes you couldn't foresee, matter you never chose, resonances that crept up around you to wait for your next move. This is not you the writer of poems. This is you the poem, this is you in the language. Not you, you in the language. Not you today, you in time.
>
> The thing you are, at that point, simply knows more than you do.[11]

Every writer needs to get out of their own way and let 'the thing you are' have free rein. This is what enables writing to serve as a tool for 'growth, change, and transformation'[12] in the workplace and beyond. The teaching and the learning happen within the writing process.

Another technique you can use is the 'loop method' of combining personal and professional writing, devised by Peter Elbow. He likens this to 'an elliptical orbiting voyage'[13] rather than the one-way journey of writing your way to a point you want to reach, and asserts that it offers the best compromise between the control of fully planned or outlined writing and the creativity in working it out as you go.[14] Elbow also suggests that his loop method is particularly useful for writing you are bored by or don't want to do. He divides the process into 'the voyage out' and 'the voyage home'.[15] The voyage out involves lots of broadly directed freewriting, where most of the creativity happens. The voyage home involves the application of control to formulate your ideas into a well-crafted piece of writing.

Elbow suggests you pick one or more categories for your directed freewriting from the following options:[16]

1. First thoughts – idea generation.
2. Biases and prejudices – related to your writing topic.
3. Instant version – a no-holds-barred attempt to create a full draft of your writing in one go, as close to the final version as possible though with some content invented if necessary.
4. Dialogues – between any conflicting feelings you may have, or between different public or historical figures, or people known to you.
5. Narrative thinking – the story of your thinking on this topic and project to date.
6. Stories – think of and briefly record all the stories connected with your topic, however tenuously.
7. Scenes – freewrite about specific moments in time.
8. Portraits – thumbnail sketches of people connected with your topic, however tenuously.
9. Alternative audience – write your piece for a very different audience than the one it is actually intended to reach.
10. Alternative writer – write as if you were someone very different from yourself.
11. Mistakes – write down everything you can think of, connected with your topic, that you know not to be true.
12. Lies – write all the weird, surreal things you can come up with around your topic.

You will probably have a clear idea about which of these categories is likely to be most useful for you. Most of them can be done in under fifteen minutes unless you are an especially slow writer or have a great deal to say. And of course if time is tight, you can use the 'write as many as I can in X minutes' approach. Remember not to censor your thoughts or edit your words as you write. At times this work may feel stupid or irrelevant, but that is normal on the pathway to new thinking. Directed freewriting helps you generate ideas, questions, insights, and energy for your work, as well as promoting creativity and productivity. Also, because this is the loop method, you can revisit the categories later and choose new ones. It is worth trying them all at some point, even – or perhaps especially – any that you find unappealing.

Begin the voyage home with a good think about your actual piece of writing: your topic, your audience, what you want to achieve. If you have written instructions or guidance for your writing task, review these carefully. When you have it all clear in your mind, read back through your work from the voyage out and gather together any useful material. There will inevitably be quite a lot of wastage, and this can feel frustrating, but every word written is a step on the journey to your finished piece of writing. Many journeys have boring, uncomfortable phases, and the journey of writing is no exception. When you have collated all the useful material, it may look like a draft of the writing you have to do – or it may not. If it doesn't, then draw on the techniques outlined elsewhere in this book to help you pull your draft together. Within that process, if necessary, you can do more loops with more directed freewriting. This might be to generate more ideas for a section where they are thin on the ground, or for further exploration of insights or lines of thought.

Writing as therapist

We have known for over eighty years,[17] and probably longer, that writing can have a therapeutic function. Writing has proved effective in actual therapy, such as for sufferers of post-traumatic stress disorder.[18] But you don't have to be in therapy for writing to have therapeutic benefits.[19] Like a living therapist, writing helps us to order our thoughts and gain greater understanding of our feelings, and it doesn't make judgements. This is not to say that writing can replace human therapists. If you are suffering from clinical depression or severe anxiety or a very low sense of self-worth or suchlike, you need more than just writing to help you recover. Nevertheless, there are several ways in which writing can be therapeutic in workplace contexts, such as reflective journalling, cathartic writing, and poetry construction.

Reflective journalling is widely used in professions which are particularly emotionally demanding, such as nursing, teaching, and social work. It is known to be a therapeutic practice, and also to enable learning and support continuous professional development.[20] That would suggest that it could be useful far beyond the 'helping professions'.

Some people are resistant to the idea of reflective journalling. If they have a go, though, they usually find that it is beneficial.[21] If you are a reluctant journal writer it may help to understand what some of the benefits can be.

One reason reflective journalling is useful in the workplace is that it externalises thought. Thinking is an essential creative process for any line of work but is often not valued by managers because they can't see what you're doing. If you are writing in your reflective journal, others can see that you are working, and this visibility makes the task more acceptable.

Another reason reflective journalling can be useful is that it enables you to track progress – both your own, and that of a project or projects. When you are working on a project over months or years, this can be helpful in many ways. It could be used to document achievements to support verbal or written reporting, keep a record of troublesome interactions in case of need, or demonstrate your skills development for an appraisal.

Reflective journals do not only have to be written in factual prose. Writing letters in your reflective journal – or even writing it entirely as letters – may 'help you think at a deeper level and give you some different perspectives'.[22] And journals may be partly or wholly written in fictional stories, poetry, cartoons, or using any other format that works for the writer.[23] People have understood the therapeutic properties of storytelling and poetry in general for millennia,[24] and reflective journalling can tap into these qualities.

Reflective journals may also help by externalising emotion, perhaps through the use of cathartic writing, which may be used within a journal or on its own. Cathartic writing is a way of expressing emotion onto the page or screen – and so it is sometimes called 'expressive writing'. The usefulness of this first became apparent in the 1980s with the work of psychologist James Pennebaker and his colleagues in the US. They researched the link between cathartic writing and recovery from trauma, and found – among other things – that cathartic writing significantly reduced absences from work.[25]

The word 'cathartic' is from the Greek for purge/cleanse/purify. Cathartic writing can purge troublesome emotions and render them more understandable and manageable. The act of putting your feelings down in words on a page or screen serves to give you a little distance from those feelings – enough to think about them as separate from yourself, even though in fact they are not.[26] This may sound like a comparatively new practice, but in fact it has been around for almost as long as writing itself.[27] Other forms of writing may be cathartic, such as memoir, autoethnography, and poetry – but here the catharsis is a by-product rather than the end goal.

Cathartic writing is similar to freewriting in that the process is simply to write without censoring or filtering what you write. The difference is that in cathartic writing, you write about your emotions. Nothing is off limits. Your feelings, reactions, desires, and wishes can be as socially unacceptable as you need them to be in the privacy of your own page or screen. You can always delete or destroy them afterwards; burning your writing can be quite cathartic in itself (though if you decide to do that please take all necessary precautions to burn it safely).

In the workplace, cathartic writing is particularly useful in two key contexts. The first is where something is generating intense emotions which are making it difficult to concentrate on work. These emotions may be negative or positive; cathartic writing can be as useful when you are deliriously happy, maybe as a result of an offer of promotion or marriage, as when you are upset or angry. The second is where a workplace encounter has not gone as expected, perhaps again leaving you upset or angry. Cathartic writing can help you to understand and regulate intense emotions and so give your work the priority it deserves.

Writing as friend

Writing is a resource which, like a very good friend, is always available. Of course, like most friends, it can be annoying or disappointing at times. Even so, it is worth finding a way to be friends with your writing because the process will be easier and the results more satisfying.

It is easy to see writing as a chore that must be done – and for sure sometimes it is exactly that. But we all use words to communicate. Even many non-speaking people use words: through devices such as a text-to-talk communicator or a spelling board, or simply through writing. People whose hearing or speech is impaired may use sign language which is also based on words. And people who can speak and hear clearly use words to communicate through speech. Writing is an extension of this process, enabling asynchronous communication through messages, emails, letters, etc., as we saw in the previous chapter, as well as more extended communications such as reports and books. Communication is essential for all of us, in our workplaces and elsewhere. So it makes sense for each of us to find out how best we can communicate in writing.

Some people hate writing, or struggle with it in a combative, adversarial way. If writing is something you have to do for work, hating it or struggling with it won't help. For sure writing can be difficult, particularly if you face extra obstacles such as writing in an additional language or being dyslexic. Yet people can write effectively and well even in the face of such problems. The Russian writer Vladimir Nabokov wrote nine novels in English; the Irish writer Samuel Beckett wrote many works in French; and the French Canadian writer Jack Kerouac, who grew up speaking French, wrote books and poems in English.[28] Well-known dyslexic writers include Benjamin Zephaniah, Irvine Welsh, and John Irving,[29] all highly successful and well-regarded authors.

So, if you're not already friends with writing, find a way to make friends. What do you like about writing? Perhaps you currently enjoy writing more as a consumer than as a producer: maybe you like reading magazines, or watching movies, or listening to podcasts. If so, start trying to notice the writing with a critical eye. What does the writing do that gives you pleasure? Why do you like that? What does it do that you don't like? Why don't you like that? Then, crucially: how could you work towards using your writing to give others similarly pleasurable experiences?

Of course, not everyone can write a great novel or screenplay or comic. But even if what you need to write is much more prosaic – a report for work, an assignment for college – you can still take a creative approach as you interact with the words you use. This can be similar to the creative approaches we take to interactions with other people. We all co-create our friendships, developing and strengthening them over time. And we can do this for our relationship with writing, too.

This is easier if you like writing in the first place. Even if you don't, you can get to like it – or at least to dislike it less.

Diaries in the workplace

One personal type of writing that has great workplace potential, but which is sometimes overlooked, is the diary. As so often, the terminology here is inconsistent. Diaries, journals, field notes, case notes, logs – these and other related categories overlap and so can be confusing. We have already discussed the therapeutic potential of a reflective journal, but there are many other uses for, and types of, diaries. These include research journals, travel diaries, scrapbooks, observation journals, personal diaries, and others.[30] Essentially, though, they are all methods of recording phenomena.

The conventional view of a diary is as something personal and private. Workplace diaries can also be personal and private, such as a reflective journal. As well as being therapeutic, this also enables you to reflect, in writing, on your work; its impact on others, and on you; what has gone well; what you would do differently another time, and why. This creates a useful record to help you assess your performance and your progress at work. Furthermore, there is a body of evidence which shows that giving a little time each day or each week to keeping a reflective workplace journal can improve professional performance.[31] There is another body of evidence which shows that keeping a diary has a positive effect on wellbeing.[32] In the light of these findings, it seems obvious that using diaries in the workplace will be beneficial.

Some workplaces encourage people to keep a diary regularly, such as daily or weekly, while others recommend more flexibility, such as writing a diary entry for any unusual occurrences. Some suggest structuring your diary with a set of relevant questions, such as: What happened? What went well? What could have gone better? What did I learn? What might I do differently another time? Others choose an unstructured approach, leaving you to choose how you organise your diary.

Workplace diaries are not always called 'diaries': they may be known as handover notes, or case files, or some other name. When I was a social worker in the last century, working with young people in residential care, at the end of each shift we wrote brief notes of each resident's activities and any problems that had arisen or achievements they had made. These were to inform the workers on the next shift, but over time they also formed a diary of each resident's progress through the care system.

There is very little information available about how to write a workplace diary.[33] This may seem problematic, but on the plus side it means you can keep such a diary in whatever form and at whatever frequency works best for you. Conventionally, diaries are written in prose, but they can also include – or be made entirely from – drawing, recorded speech, photos, emails, poems, videos, even stitching, or collage, or scrapbooking.[34] They may be created digitally or physically.

The received wisdom is that making regular entries is ideal. Some of us, though – including me – are ad hoc diary writers at best. Yet I have a kind of 'distributed diary' or 'embedded diary' which I can use to track back my activities and thoughts if I want or need to do so. This is made up of my ad hoc diary writing plus emails, my electronic calendar, spreadsheets I use to keep records of tasks and time spent, notes of meetings, my blog, and so on.

Conclusion

When you are writing purely for your own purposes and not for other readers, you are free to be as creative as you like. This means your writing can be as messy, biased, impartial, polemical, systematic, or chaotic as you wish. It is well worth daring to be exploratory in your personal writing because you will learn about what is possible and what seems to work. Then that learning will transfer over to the other kinds of writing you do.

A life-changing experience

In July 2019 I was asked to help with a conference at the University of Kent. I hadn't worked at that university before, and was happy to find a verdant campus with trees and rabbits on a hill overlooking the historic city of Canterbury. The conference was over two days: the first on autism and participatory research, the second on autism and gender. Around fifty people attended each day, most of them autistic.

The conference was wonderfully inclusive. Conference bags held fidget spinners, ear plugs, and stickers in green and red to indicate 'I'm happy to chat but I might struggle to initiate conversation' and 'I probably don't want to talk to anyone right now'. There were rooms set aside for people who wanted time out, and the hand dryers were turned off in the toilets with paper hand towels provided instead. People could, and did, dip in and out of sessions to suit their needs. During sessions people made themselves comfortable in any way they chose: sitting on chairs or on the floor, lying on the floor, facing towards the presenter or away, moving or being still.

At the start of the first day, the helpers came out to the front and each of us introduced ourselves and our role. I said something like 'I think I'm neurotypical, although one of my autistic friends tells me that I'm neither neurotypical nor autistic, so probably I'm in a category that doesn't yet have a label.' By the end of the first day, several people had come up to me, independently of each other, to suggest that I might need to rethink my friend's assessment. I learned that someone at the conference had a form of synaesthesia that enabled them to distinguish between neurotypical and autistic people. The next morning I was introduced to this person, who took one look at me and said, 'You're a lime green pencil. You're definitely autistic.' Apparently they see autistic people as definite shapes and colours, while neurotypical people look spiky and fuzzy.

This was a shock and it took me a while to process. I had less to do in the second day of the conference, so I decided to sit and listen to the presentations for a while. I learned about how difficult it can be for women and girls to get a diagnosis, because the diagnostic criteria were mainly built around how some boys manifest autism. I thought maybe it wasn't worth bothering to go for an assessment myself. Then I learned about the 'lost generation' of autistic people, mostly women, who were not diagnosed in childhood because their manifestations of autism did not fit the diagnostic criteria. I began to wonder whether I was one of them, and decided I wanted to know.

I took the AQ10 test, which is clinically validated, and scored nine out of ten. I took that to my doctor who was kind, understanding, and knowledgeable. He sent me a fifty-question test which I completed and returned, though some of the questions seemed out of date (Q: Do you remember all the phone numbers you use? A: I used to when phones were static, but now I outsource my memory to my mobile phone.)

Then the pandemic arrived and slowed everything down. A year later I got a letter saying my first appointment had been scheduled with the Adult ADHD and Autism Service of my local NHS Healthcare Trust. The first stage was a phone appointment lasting an hour, in which I had to answer a long list of questions about my childhood, upbringing, and interpersonal experiences. The psychologist asking the questions was very kind, but it was exhausting. Then the Service sent me a form for 'someone close to me' to fill in, which asked about things like my ability to engage socially, whether I take things literally, and my response to changes of plan. Once that was done, I had a two-hour video appointment with a mental health nurse. He told me the assessment was multi-disciplinary and that, although in this appointment I would only see him, he had consulted with his colleagues. Then he asked me another whole bunch of questions, some of which I had already been asked by the psychologist on the phone, and finally told me I am autistic. 'There is no doubt', he said.

Although I had a long wait for assessment, and by this point I had already concluded that I was autistic, the diagnosis still came as a shock. But it also explained a whole lot of things I had spent my life being unable to understand. I hadn't told many people that I was going through the assessment process, and I decided to come out to my family, friends, and colleagues. I told some people in person, others via message or email, and when I had done enough of that, I put it on social media. I was extremely lucky because almost everyone responded positively and supportively. Several people said things like 'I don't know anything about autism but I know you and I love you and please tell me all about it when we speak', which was fine by me. Many were genuinely interested, which was nice. Some colleagues floored me with the extent of their support and care.

Now, some years on, I still feel as if I'm getting used to being autistic. Which is odd, because of course I have been autistic all my life. But life as a knowingly, openly autistic person is very different from life as an undiagnosed autistic person. And, I'm glad to say, it's a lot better.

Try it yourself

I am offering you two options here. I suggest you choose whichever one appeals to you, though you can do both if you like – or do one now and return to the other later on.

Option 1: freewriting. Do five minutes of freewriting about something you feel strongly about. Choose a prompt first, and remember it must be an active prompt, such as 'I am angry

about the political situation in my home country because ... '
or 'I feel upset with my colleague because ... ' Then set a timer
for five minutes (or whatever short time period you prefer)
and write, without stopping or editing, for that length of time.
When you have finished, read over what you wrote. Did you
learn anything? Did anything you wrote surprise you? How do
you feel about the issue now? Is there anything else worthy of
note? Write down the answers to these questions, and reflect
for a few minutes on your answers, the text you produced, and
your freewriting experience to see if there are any useful elements you can take forward.

Option 2: structured letter. Write a letter to a person in your workplace who has more power than you: perhaps your manager, tutor, or supervisor. This is not a letter intended for sending, though you can, if you like, when it's done. The purpose of writing the letter is to help you gain more clarity about your relationship with the person. Use this structure:

Dear [whatever salutation you would normally use]
> I am grateful to you for ...
> I am angry/frustrated with you because ...
> What I really want from our working relationship is ...

Then sign off as you choose. Write as little or as much as you like, but don't spend too long on it and, crucially, don't overthink it. Remember it is for your own purposes, so treat it a bit like freewriting and focus on getting the writing done. When you have finished, read over what you wrote and give yourself a little time for reflection. Did you learn anything? Did anything you wrote surprise you? How do you feel about the relationship now? Is there anything else worthy of note? Write down the answers to these questions, and reflect for a few minutes on your answers, the text you produced, and your experience of writing the letter to see if there are any useful elements you can take forward.

Notes

1 Helen Sword, *Writing with Pleasure* (Princeton University Press, 2023), p. 7.

2. Paul Silvia, *How to Write a Lot: A Practical Guide to Productive Academic Writing* (American Psychological Association, 2007), pp. 3–4.
3. Sword, *Writing with Pleasure*, p. 8.
4. Richard Phillips and Helen Kara, *Creative Writing for Social Research: A Practical Guide* (Policy Press, 2021), p. 17.
5. Peter Elbow, *Writing with Power: Techniques for Mastering the Writing Process* (2nd edn) (Oxford University Press, 1998), pp. 13–19.
6. Robert King, Philip Neilsen, and Emma White, 'Creative writing in recovery from severe mental illness', *International Journal of Mental Health Nursing*, 22.5 (2013), pp. 444–452.
7. Sally O'Reilly and Linda Anderson, 'Stimulating creativity and imagination', in *Creative Writing: A Workbook with Readings*, ed. by Sally O'Reilly and Jane Yeh (2nd edn) (Routledge, 2022), pp. 13–34.
8. Clare Morgan, Kirsten Lange, and Ted Buswick, *What Poetry Brings to Business* (University of Michigan Press, 2010), p. 52.
9. Elbow, *Writing with Power*, p. 42.
10. *Ibid.*
11. Gavin Maxwell, *On Poetry* (Bloomsbury Academic, 2012), pp. 118–119.
12. Rich Furman, 'Beyond the literary uses of poetry: A class for university freshmen', in *Poetry as Therapy, Research, and Education: Selected Works of Rich Furman*, ed. by Rich Furman (University Professors' Press, 2022), pp. 241–248 (p. 241).
13. Elbow, *Writing with Power*, p. 60.
14. *Ibid.*, p. 59.
15. *Ibid.*, pp. 60–77.
16. *Ibid.*, pp. 61–73.
17. Natalie Robinson Cole, 'Creative writing as therapy: Or, nobody's an angel', *The Elementary English Review*, 20.1 (1943), pp. 2–6.
18. Ellen Barry, 'A novel therapy, using writing, shows promise for PTSD', *New York Times*, 23 August 2023, www.nytimes.com/2023/08/23/health/ptsd-writing-therapy.html [accessed 1 September 2023].
19. Ana Caterina Costa and Manuel Viegas Abreu, 'Expressive and creative writing in the therapeutic context: From the different concepts to the development of writing therapy programs', *Psychologica*, 61.1 (2018), pp. 69–86 (p. 72).
20. David Bryson, 'Continuing professional development and journaling', *Journal of Visual Communication in Medicine*, 44.4 (2021), pp. 198–200, https://doi.org/10.1080/17453054.2021.1974292

21 Lorna M. Dreyer, 'Reflective journaling: A tool for teacher professional development', *Africa Education Review*, 12.2 (2015), pp. 331–344, https://doi.org/10.1080/18146627.2015.1108011
22 Barbara Bassot, *The Reflective Journal* (3rd edn) (Bloomsbury Academic, 2020), p. 110.
23 Nicole Brown, *Making the Most of Your Research Journal* (Policy Press, 2021), p. 47.
24 Rich Furman, 'Poetry therapy as a tool for strengths-based practice', in *Poetry as Therapy*, ed. by Furman, pp. 3–18 (p. 4).
25 Martha E. Francis and James W. Pennebaker, 'Putting stress into words: The impact of writing on physiological, absentee, and self-reported emotional well-being measures', *American Journal of Health Promotion*, 6.4 (1992), pp. 280–287.
26 Michael Rosen, 'In these troubled times we all get the "bothers" but I have a surefire cure: write them down', *The Guardian*, 15 September 2023, www.theguardian.com/commentisfree/2023/sep/15/the-bothers-cure-write-sentence-school-michael-rosen [accessed 11 November 2023].
27 Claire Williamson and Jeannie Wright, 'How creative does writing have to be in order to be therapeutic? A dialogue on the practice and research of writing to recover and survive', *Journal of Poetry Therapy*, 31.2 (2018), pp. 113–123 (p. 114), https://doi.10.1080/08893675.2018.1448951
28 Cedric Lizotte, 'Prolific polyglots: Authors who wrote in a second language', *The Airship* (undated), http://airshipdaily.com/blog/6232014prolific-polyglots-authors-who-wrote-in-their-second-language [accessed 13 November 2023].
29 Manpreet Singh, '10 famous writers who once struggled with dyslexia', *Number Dyslexia*, 11 February 2022, https://numberdyslexia.com/famous-writers-with-dyslexia/ [accessed 13 November 2023].
30 Sword, *Writing with Pleasure*, pp. 146–148.
31 Anselmus Sudirman, Adria Vitalya Gemilang, and Thadius Marhendra Adi Kristanto, 'Harnessing the power of reflective journal writing in global contexts: a systematic literature review', *International Journal of Learning, Teaching and Educational Research*, 20.12 (2021), pp. 174–194 (p. 174), https://doi.org/10.26803/ijlter.20.12.11
32 Janet Salmons, 'Journaling right and left', in *Creative Expression and Wellbeing in Higher Education: Making and Movement as Mindful Moments of Self-Care*, ed. by Narelle Lemon (Routledge, 2023), pp. 33–52 (pp. 37–41).
33 Brown, *Making the Most of Your Research Journal*, p. 3.
34 Helen Kara, *Qualitative Research for Quantitative Researchers* (SAGE, 2022), p. 110.

8
Good practice in writing

Introduction

We often think of language as fixed, but in fact it is mutable and ever-changing.[1] Grammar, though, i.e., the rules of language, shifts much more slowly and infrequently, if at all. Some people learn grammar instinctively as they learn a language while others need to learn the rules.

There are a whole load of grammatical rules covering all aspects of writing in practice, from word choice to sentence and paragraph structure, verbs and tenses, adjectives and adverbs, nouns and pronouns, and many more. It is beyond the scope of this book to cover them all in detail. It is useful to know that the grammar of speech is different from the grammar of writing – or at least different from prose writing; it is more akin to poetry.[2] This explains why, when you transcribe a quote verbatim into an assignment or report, it can read strangely, perhaps seeming repetitive or disjointed, though it may have sounded entirely articulate in speech.[3] (There is an example of this in Chapter 3.)

Different languages have different rules. Let's look at a simple example. In English (and other languages such as Portuguese, Bulgarian, Chinese, and Swahili) sentence order is subject–verb–object, or SVO. So you would say: Helen wrote a book. ('Helen' is the subject of the sentence, 'wrote' is the verb, and 'book' is the object.) In some other languages, such as Korean, Turkish, and Punjabi, the order is SOV: Helen a book wrote. Then in languages such as Arabic it is VSO: Wrote Helen book.[4] So it is easier to learn a new language if it has similar grammar to a language you already speak.

One of the joys of writing is finding the word, phrase, or sentence which seems adequately to describe a place, a sensation, a moment. When we are writing, the only way we know we have found that word, phrase, or sentence, is that it *feels right* to us. This makes writing a slow process at times. We have to stop and savour our own words – not every time, or all the time; when words are pouring out in torrents, the savouring can come later. But we need to slow down, periodically, as we write, to check whether our words are working for us.[5] If you write by hand, the only real option is to read your work out loud. If you write digitally, you can also read out loud, or print your work to read instead of reading it on screen (or read it on screen, if you usually read print-outs), and/or change the font and/or the background colour of the screen or paper. These tricks help to give us enough distance from the text we have created to be able to experience it afresh, and consider, again: do our words work?

Myths about writing

There are eight myths about writing which I would like to debunk. These are:

1. I can't start writing until I know what I want to write down.
2. I have to be in the right mood to write.
3. I can't write unless I have at least half a day free.
4. I need to start writing at the beginning of the piece.
5. I must write it right first time.
6. Everything I write is rubbish.
7. My writing will never be good enough.
8. Writing is too hard for me; I can't do it.

Let's take these in turn. If you read the previous chapter, you will know that we learn from our writing. If you're not sure what or how you want to write, techniques such as freewriting and reflective journaling can help. Details of these techniques are also in Chapter 7, and they can help you to start writing, even if you don't know what you want to write.

Writing is part of your job. You show up for work when it's time to show up for work, not when you're in the right mood for work. Same with writing. Show up and write.

You can start writing anywhere you like. In the middle, at the end – doesn't matter. Perhaps in the old days, when longhand was the only option, or typing with no delete key, you would have had to start at the beginning and write through to the end. These days, working digitally, it is so easy to move chunks of writing around that there is no need to worry about the structure of your work too much until you have the content. Right now, as I write this sentence, this chapter has fewer than a thousand words in three chunks. I have a few paragraphs which I think will be the introduction (though that may change), one paragraph about finding the right word or phrase, and this section. There is only one sub-heading, right at the start ('introduction'), and I don't know what order these chunks will end up in. It doesn't matter. I know I can fix that later.

It is really very difficult to write something right the first time. It is not impossible, though the longer and more complicated the piece of writing you need to do, the harder it will be. As we saw in the last chapter, Peter Elbow describes trying to write something right the first time as 'the dangerous method'.[6] He warns that, unless you are a skilled professional writer, trying to write this way can end up being more time-consuming than writing a rough draft then editing and polishing. I am an experienced author with a number of books to my name and I think 'the dangerous method' would certainly be more time-consuming for me. There is a helpful sense of freedom in allowing my words to flow unhindered and trusting that, at a later stage, I will be able to craft the result into something worth reading.

Almost all of us have critical voices in our heads telling us our writing isn't any good. I have them, despite the number of books I have published and all their good reviews. Our inner critics can obstruct the writing process, so find a way to deal with them. Visualisation can be useful unless you're aphantasic. I use a visualisation which works for me: my inner critics appear in my mind's eye as evil little goblins, which I get rid of by bashing them into outer space with a frying pan. I don't know why a frying pan works so well – I am not given to using frying pans or indeed anything else as a weapon – but in my mind it is pleasingly cartoon-like and effective. A friend of mine visualises a party in the back of her head, tells her inner critics, 'Come on in, the party's in the back', and sends them off to have fun there while she gets on with her work. If you are aphantasic, visualisation won't work for you, so I suggest coming up with a very dismissive phrase to use whenever your inner critics make their presence felt. As you will only be using it in the privacy of your own brain, it can be as rude as you like.

Your writing will be good enough if you keep working at the craft. An old saying is that 'practice makes perfect', but that's not really applicable to writing because very little writing is perfect. Creating new editions of books really drives this message home. Publishers don't want new editions of books unless those books have sold well, so you're working on a book which is already good. Then you pull it to pieces and put it back together, with some new sections and lots of revision, to make it a better book. I suspect the only examples of perfect writing are all very short. But although, with writing, practice does not make perfect, practice will certainly make your writing better.

If writing feels hard to you, that's not because you can't write, or because of any lack in you. It's because writing IS hard. That said, it's not so hard you can't do it. There are some people for whom writing is too hard, such as people who have certain forms of severe dyslexia, or learning disabilities, or cognitive impairments. It is unlikely that people for whom writing is too hard will be reading this book. You are reading this book because you are interested in writing and you want to be able to do it better. You can do it better, and you will. Think of all the writing you have done to date: emails, social media updates, reports, assignments, lists, blog posts, and no doubt lots more besides. You have already done a great deal of successful writing. I realise that may not stop you feeling daunted by your current writing project, but nevertheless it is worth bearing in mind.

Productivity

If you are writing for work – or for work-related study – then writing is, quite literally, your job. OK, it's probably not the whole of your job, but still, writing is something you need to show up and do, just as you need to show up and do your job in general. Yet this is often easier to say than it is to do because of the enemies of productivity. These include: perfectionism, misconceptions, critical inner voices, and distraction.

Good practice in writing

As we have seen, there is rarely such a thing as a perfect piece of writing. Maybe a sentence, or a short poem, but beyond that there is always scope for revision and improvement. Perfectionism can be useful at final draft stage as long as it is not too extreme. But it has no place in the first draft, and not much place in the second. In a professional environment where we are expected to show high levels of competence it can be difficult to relax enough to write a useful first draft. It feels a bit like that old childhood game of trying to pat your head with one hand and rub your tummy with the other. You need to be competent in the workplace and also willing to write what may at first seem like a load of rubbish. But, as with the game, if you practice you will develop the abilities you need.

The main misconception about writing is the belief that if you have a big piece of writing to do, you need big chunks of time for that writing. It may seem counter-intuitive, but you will be more productive in regular short periods. In a focused half-hour most people can write somewhere between two hundred and eight hundred words. You will write less on a slow day or if you are writing in an additional language, more when the words are flowing and in a language in which you are fluent. I am writing this book at fifteen hundred words a week, which is nominally three hundred words a day Monday to Friday, though in practice my schedule is irregular as I fit my writing in around other commitments.

Books on writing productivity tell writers to set a schedule, with goals, and monitor that schedule to make sure they stay on track. I use a spreadsheet for my schedule. Table 8.1 shows the spreadsheet exactly as it was when I had written the last sentence. You can see that I have done some work on every chapter except the last, though only the first three chapters are finished at this point (and one of those still needs a structural amendment). You can probably also tell that I use word goals rather than time goals. I am more productive with word goals: if I have five hundred words to write, I may well do that in under half an hour; if I set myself half an hour to write, I am more susceptible to inner voices and distraction (more of them in a moment) and so likely to write less than five hundred words. But everyone is different and you need to find out what works best for you. If you don't know, try word goals and time goals and you will soon figure it out.

Table 8.1 Spreadsheet

Chapter	Aiming for	Words written	Plus		Status
1 - introduction	6000	6404			Done
2 - stories and fiction	6000	5299	658	in story	Done for now. Need to move sensory language bit up
3 - writing from life	6000	6154	489	in memoir piece (done)	Done
4 - scripts and screenplays	6000	1112			
5 - cartoons, comics and zines	6000	4445	137	in zine	
6 - poetry	6000	2740	54	in poem (done)	
7 - epistolary and digital writing	6000	1850	544	in letter (done)	
8 - the personal is professional	6000	3521	943	in account (done)	
9 - good practice in writing	6000	3885			
10 - conclusion	6000				
Try it yourself exercises		1195			Done for chapters 1, 2, 3, 4, 8 and 9
Totals	60000	36605	2825		

Some people like to use a system called 'the chain' which involves writing a minimum amount every day. The minimum can be really very small: just a sentence or two, if you like. Of course if you set such a low minimum you will need to do more than the minimum on most days. But it is sensible to set an achievable minimum for the days when you are sick, or someone you care for is sick, or you are preoccupied with a major life event such as moving house or changing job. Then make yourself a chart of days, which can be on paper or on screen, so you can mark off each day when you have done at least the minimum amount of writing you have set yourself. The more days you have marked, the less you will want to 'break the chain' by missing a day – because if you do miss a day, you have to discard that chart and start a whole new one. Some people find this visual reminder of their writing progress to be a powerful motivator. If you think you might be one of those people, then give it a try.

We saw earlier in the chapter that all writers have critical inner voices which say things like 'you're stupid' and 'you can't write'. Our critical inner voices are related to impostor syndrome, which is a feeling that we are not as clever or skilled as others seem to think. This is common for writers: the American novelist and poet Maya Angelou, the British entertainment journalist Nick McGrath, and other well-known writers have acknowledged experiencing impostor syndrome. I get it every time I send a final manuscript off to a publisher to go into production. Something in my brain starts telling me, 'This time they're *really* going to find out you're a fraud who knows nothing.' I find it helpful to remember that impostor syndrome is a problem of success. Reading some of the reviews of my books, or a couple of the letters and cards from my 'nice letters' file (see Chapter 6), helps to quieten the critical inner voice which is telling me I am an impostor.

The early twentieth-century thriller writer Raymond Chandler developed a technique he called 'The Nothing Alternative'.[7] This involved a self-imposed rule: he could sit at his desk and either write or do nothing. Doing nothing was boring and so his rule motivated him to write. That was easier for Chandler because he was writing before the invention of mobile phones and the internet. These days we have to make an effort to create distraction-free time for writing. That might involve using an internet blocker such as Freedom, AntiSocial, or Cold Turkey, and switching off your mobile phone. If you have a mobile phone and need to keep it on for a specific reason, such as for potential contact from your child's school or a medical professional, you can switch off notifications for any messaging apps and/or social media and only leave the notification switched on for calls.

Even then it is easy to be distracted from writing. Writing is often difficult, and sometimes it feels impossible, and then it is tempting to turn to the ironing, or the shopping, or a cupboard that needs reorganising. But this is a cop-out, an avoidance strategy, and it is important to recognise that to avoid sabotaging your own writing plans. A short preparatory ritual, such as making your favourite drink in a special cup or glass, or lighting an oil burner primed with a scent that helps you to concentrate, can be helpful for writing. If you find yourself deciding to clean all the floors in your home or rearrange your entire wardrobe before you start writing, that is procrastination which is unhelpful.

The importance of feedback

Feedback is of great value to writers, whether we are receiving feedback or giving it to others. Even so, receiving and giving feedback can be stressful, particularly for inexperienced writers. We saw in Chapter 1 that writing is a very emotional business and feedback is no exception. This is partly because it is in fact criticism – ideally constructive criticism, but still criticism of our writing, and that can be quite painful to receive.

Where possible, get feedback from one or more people who are representative of your readers before you finish your piece of writing. Feedback itself is – or should be – a job of writing. Written feedback is the most useful kind because you can take your time to digest and use the information you have received. First, give yourself time to react and deal with your feelings. You may need to jump around and whoop because the feedback is so positive, or have a cry/a hug/some chocolate because it isn't, or anything in between. Acknowledge and process your emotions before you do anything else. I recommend taking at least twenty-four hours between the time you first read the feedback and the time you start dealing with it, to help you take a more balanced view.

When you are ready to deal with the feedback, take an analytic approach. Pull out all the specific points made and sort them into three categories: the no-brainers, the no-thanks, and the oh-waits. No-brainers are points you want to implement. No-thanks are points you don't want to implement, perhaps because the person giving feedback has misunderstood your intentions, or is suggesting something you don't think is relevant, or for some other reason. Oh-waits are points you probably want to act on, not to implement as suggested, but in some other way. The person giving the feedback might make a suggestion you initially reject as a no-thanks, but then you realise that the suggestion indicates they haven't understood the point you are trying to make, so you decide to revise for clarification. When you have finished categorising the feedback, for each point write down what you intend to do (or not do) as a result and why. This provides you with an action plan for revising your work.

When someone has taken the time to give feedback on your work, it is only fair to consider their input carefully. However, remember that you are the author and as such you get to have the final say. Your record of what you decided and why is helpful if someone who gave feedback asks you why you didn't do, or did something different from, what they had suggested.

Giving other people feedback on their writing is a useful exercise for any writer. It can seem daunting at first: you may think you won't have anything to offer, but that is unlikely to prove true. Here are my top tips for giving good quality feedback:

1. Read the piece you're giving feedback on carefully and thoroughly, at least twice.
2. While you read, make notes of thoughts that occur to you. As a minimum, these should include: aspects of the work you think are good; where you think there is room for improvement; anything you don't understand.
3. Be sure to praise the good points in the author's work. This helps them to trust your feedback and improves their confidence.
4. Be open about anything you don't understand. Doing this worries some people because they think they may look stupid, but it's really helpful feedback for writers because it can indicate that they haven't written clearly enough.
5. Give a straightforward assessment of areas where you think there is room for improvement.

6. Tell the author *how* you think they can improve their work. This is crucial. If you're only saying *where* improvement is needed, you're only doing part of the job.
7. Be honest in all the feedback you give.
8. Don't worry if you can only offer a certain amount of help because of the limits to your own knowledge. It's fine to say, for example, that a quick online search suggests there is more information in the area of X; you're not certain because X lies outside your own areas of expertise but you think it would be worth the author taking a look.
9. Acknowledge the author's emotions. For example, after giving quite critical feedback, you might say something like, 'I realise that implementing my suggestions will involve a fair amount of extra work and this may seem discouraging. I hope you won't be put off because I do think you have a solid basis here and you are evidently capable of producing an excellent piece of writing.' (Though remember point 7 above and don't say this if it's not true.)

You may receive feedback which isn't up to these standards, such as feedback which tells you what needs to be improved without saying how those improvements should be made. If so, and if possible, go back to the person and ask them for more input. Explain what you need from them. If it would be helpful, point them to this section; sometimes it can be useful to refer to an external authority.

Who are you writing for?

Who are your readers or audiences? Do you know? If not, can you imagine who they might be?

Figuring out who you are writing for can help you to write more easily. You may know exactly who you are writing for: your manager, tutor, team, or the like. Or you may be writing for an unknown readership, in which case try to imagine who might be a typical reader. The reader in my mind as I write this book is interested in improving their writing, is working and studying, is intelligent and curious, enjoys learning, and has a good sense of humour. Of course not all my readers will be exactly like this – some may be retired, or serious, or more interested in the theory of writing than in its practice – but holding the image of this reader in mind helps me to find a suitable voice to communicate with them. Some writers like to find a picture on the internet of a person who looks like their imagined reader, print it out and stick it up on the wall by their desk to help them hold that person in mind.

Knowing who you are writing for helps you to figure out how to tell your story in a way that is likely to appeal to your readers or viewers. Sometimes you need to tell the same story in different ways for different audiences. This requires some creativity and it can also be a lot of fun. Many years ago I did some research with an organisation supporting families with pre-school children. The organisation's funders wanted me to write a conventional research report; fine, I could do that. The families asked me to write a report they could read with their young children at bedtime. That was a wonderful task to be given. After considerable thought, and a couple of false starts, I created a photo-essay. Each research finding was written in plain English with an illustrative photo and quote from the data. I am glad to say the families loved their report.

Writing in an additional language

Learning to write to the standard required for professional purposes is challenging for a native speaker. If you are writing in a language which is not your first, it will be more challenging still.[8] We saw earlier in this chapter that different languages have different grammatical rules and that the grammar of speech is different from the grammar of writing. Also, there is much more pressure on correct grammar in writing than in speech. Speaking a new language is easier to learn than writing because we can reach a level of adequate communication without being fluent, and native speakers are usually ready to forgive any errors and focus on the meaning we are trying to convey. In workplace writing this is not the case and a high standard will be expected.

Some people choose to write in their first language and then translate what they have written into the language of their workplace. Everyone has the right to do this if they wish. However, in general it is not a great idea for two main reasons. First, it is a much slower route to learning to write fluently at a high standard in the additional language. Second, translation is not a simple straightforward process. Idiomatic phrases, metaphors, sometimes even single words may have no direct translation. Words and expressions can have different meanings in different cultures: for example, red often stands for danger in the UK, but in China it is the colour of good luck. Expressive nuances of tone, such as humour or irony, can be particularly difficult to translate. If you get stuck on a word and use the relevant word from another language for the time being, that's fine,[9] but doing as much as you can in your workplace language will save you time and improve your learning.

It is easy to focus on what we can't do or what we are struggling with, and forget how far we have already come. If you have secured a job where the primary language is not your first language, then you have already reached a high level of proficiency in that additional language. Granted, that may be due more to verbal than written proficiency – but, if you practice and persevere, written proficiency will follow.

Whatever your work may be, it is likely to have a specific vocabulary which helps colleagues to communicate with each other about the work they are doing. Medicine, law, engineering, catering, theatre, shipping, and so on all have different vocabularies. To make life even more complicated, some words may refer to different things in different professions. In the construction industry, 'scaffolding' refers to metal pipes which are used as a temporary structure to assist with creating, maintaining, and repairing the built environment. In education, 'scaffolding' refers to the support and guidance a teacher will give to students to help them learn new skills. In engineering, 'materials' refers to cement, steel, glass, and so on, whereas in fashion design 'materials' refers to things like textiles and accessories such as buttons and sequins. There are many other such examples.

Everyone joining a new profession has to learn its specialist vocabulary. This can be challenging even for people working in their first language, and more so if you're working in an additional language. For people outside the profession, its specialist vocabulary is known as jargon, but for insiders it is an invaluable way to communicate. This means you need to know when to use the specialist vocabulary of your profession – essentially, when you are addressing other members of that profession – and when to convey the same messages in plain language. If in doubt, use plain language. People who don't understand jargon may find it confusing or offensive, while people who do understand jargon are very unlikely to be confused or offended by the use of plain language.

When to be a little less creative

There are some professional contexts where creativity in writing is more restricted, usually where any ambiguity is unwelcome. Examples include writing up information sheets to go with medication, instructions for an action, or the analysis of intelligence in the criminal justice system. For the latter, the UK College of Policing gives very clear guidance about how to write intelligence analysis for impact on decision makers. Among other things, they suggest using the ABC method and the 4–3–3 principle.[10] ABC stands for accuracy, brevity, and clarity. The 4–3–3 principle states that no sentence should be longer than four lines, no paragraph should be longer than three sentences, and no section should have more than three paragraphs. (I find the 'four lines' guidance unhelpful, because the length of a 'line' depends on whether the writing is by hand or digital, and on the size of the handwriting or font. For me, the Plain English Campaign guidance that sentences should vary in length with an average of fifteen to twenty words per sentence[11] is more useful.)

Even though the UK College of Policing is so prescriptive about how its members should write their analytic work, it recognises the potential role of creativity in illustrations for that work. These can include, among other things, graphs, tables, pictures, maps, and infographics – which last can 'provide a high level of creative licence to the analyst'.[12] So although the College does not want its members to be too creative as they write their text, evidently they encourage members to be creative with visual elements.

As we learned in Chapter 4, the boundary between drawing and writing is blurred. Good infographics 'explain things graphically',[13] using visual images such as charts, illustrations, photos, and diagrams, as well as some text. The text in a good infographic plays a supporting role, and so should be kept to a minimum, with most of the story told through the visual elements. There are plenty of examples of good and bad infographics online. Initially infographics were printed images, but they now come in digital formats such as interactive, where viewers can click to engage with the content, or animated, where the infographic is presented as an animation. And making infographics, in any format, is a creative process.[14]

Scope for creativity in writing may be limited in some workplaces, but it is not removed. You still have to choose and arrange your words, and you may well be able to use a device here and there such as tension, a metaphor, or some repetition. And you will still need to tell a story. Even the most fact-laden storytelling can be compelling when the story is told in an effective way. Think of a good quality documentary or piece of reality-based television you have seen and that will neatly illustrate my point. There are written examples, too, such as Saket Soni's book *The Great Escape: A True Story of Forced Labor and Immigrant Dreams in America* (Algonquin Books, 2024) referred to in earlier chapters. This is a non-fiction book written like a novel, and there are many others besides: memoirs, travel narratives, and the like. So even if the scope for creativity in writing is limited in your workplace, you can still create a good story and tell it effectively.

Conclusion

Good practice in writing is good practice in creative writing, because – as we saw way back at the start of this book – all writing is creative. And good practice always involves showing up and doing the job of writing, being as productive as you reasonably can, and working effectively with feedback. In workplace writing, good practice also takes into account the type of writing regarded as 'good' in your place of work. Wherever your workplace may be – office, classroom, ship's bridge, theatre, factory, woodland, or anywhere else – writing is about communication. Keep your audience in mind, write for them, and you won't go far wrong.

I interview myself about writing

HK1 Have you said it all now? Everything you know about writing?

HK2 Nowhere near.

HK1 Why not? Aren't you supposed to put everything in when you're writing a book?

HK2 Not possible. There's no way to write down everything about anything.

HK1 What do you mean?

HK2 Take this phone on my desk. I couldn't even tell you everything about that. To do so, I would need to know when it was made, where, by whom, from what; when and how it came to me, which was so long ago I can't remember; who has called and messaged me on it, when, how often; who I have called and messaged with it, when, etcetera. And that's just one artefact. So to write down everything I know about writing …

HK1 OK, I get the point. How about everything you know about learning to write?

HK2 I started with a pencil. That worked well. I learned to form my letters, and when my teacher decided I was ready, I graduated to using a pen. Then it all went wrong.

HK1 Why?

HK2 I'm left-handed. My writing turned into one long smudge until I learned to contort my hand to keep it below the line as I wrote. Nobody suggested tilting the paper instead, which I gather is how they teach left-handed children these days. But at least the teachers didn't try to make me write with my right hand.

HK1 Why would they have done that?

HK2 Because being left-handed was seen as deviant, even evil, in many countries for centuries. Including ours.

HK1 So you weren't discriminated against for being left-handed?

HK2 Not explicitly, though being left-handed frequently makes me aware that I live in a world which wasn't designed for me.

HK1 How come?

HK2 Scissors, spiral notebooks, tin openers, mugs with designs inside which, for me, are always upside-down or on the side I can't see, vegetable peelers, retractable tape measures –

again, the numbers are always upside-down – corkscrews ... then at uni we had those chairs with desk attachments which are always on the right-hand side ... and cheque books, though we don't use them now ...

HK1 Wow, I never realised ...

HK2 Then there's the terminology. The Latin for 'left' is 'sinister' which was adopted in English as a word meaning 'potential for wickedness or evil'. The French for 'left' is 'gauche' which was adopted in English to mean awkward, clumsy, and tactless.

HK1 Sounds like you.

HK2 Thanks for that. Though you're right, which reminds me of the correlation between autism and higher levels of left-handedness.[15]

HK1 Does being autistic make learning to write more difficult too?

HK2 Not for me, because I love words and writing with full autistic passion, though it may do for others.

HK1 So how did you learn to write, apart from the mechanics?

HK2 I read a lot. And thought a lot. And wrote a lot. Still do.

HK1 Is that it?

HK2 Not entirely. Some of what I read was books on how to write. And I went on some writing and storytelling courses. But essentially reading, thinking, and writing were the main ingredients of my learning, and they still are.

HK1 You mean you're still learning? Even now, when you've written so many books?

HK2 I don't think I will ever stop learning how to write. The best I can do today will not be as good as the best I can do this time next year.

HK1 So you'll always be learning?

HK2 Yes. And that's fine by me.

Try it yourself

Thanks to Glyn Maxwell[16] for inspiring this exercise. Think of an experience you remember well. Nothing too complicated; something fairly straightforward, but with a narrative element and a bit of drama. So not the last time you went swimming, but the time your swimming costume fell off in the water and sank out of reach, and how you managed to get out of the situation. Something like that.

When you have thought of a suitable experience, put it in writing in at least three of these ways – more if you like.

1. Life writing
2. Story
3. Script
4. Screenplay
5. Comic
6. Zine
7. Poem
8. Letter or email (begin with a salutation, to specify who you are writing to)
9. Reflective journal

Then compare your outputs.

- What does each reveal?
- What does it conceal?
- What does each form enable?
- What does that form constrain?
- What else have you learned from these comparisons?

Write down your answers to these questions. Keep them for reference. Repeat the exercise as needed.

Notes

1 Rachel Grenon, *Grammar: The structure of language* (Wooden Books, 2018), p. 6.
2 Ronald Carter, *Language and Creativity: The Art of Common Talk* (Routledge, 2004), p. 10.
3 Helen Kara, *Creative Research Methods: A Practical Guide* (2nd edn) (Policy Press, 2020), p. 174.

4 'Word order in different languages', *The TEFL Academy*, 8 August 2017, www.theteflacademy.com/blog/word-order-in-different-languages/ [accessed 16 August 2023].
5 Michael Rosen, 'In these troubled times we all get the "bothers" but I have a surefire cure: write them down', *The Guardian*, 15 September 2023, www.theguardian.com/commentisfree/2023/sep/15/the-bothers-cure-write-sentence-school-michael-rosen [accessed 13 October 2023].
6 Peter Elbow, *Writing with Power: Techniques for Mastering the Writing Process* (2nd edn) (Oxford University Press, 1998), p. 39.
7 Roy Baumeister and John Tierney, *Willpower: Why Self-Control Is the Secret to Success* (Allen Lane, 2012).
8 Jack Richards, 'Preface', in *Second Language Writing*, ed. by Ken Hyland (Cambridge University Press, 2003), p. xiii.
9 Katherine Firth, 'Writing a PhD in your second language: Seven reasons you're doing great and five ways to do even better', *LSE Impact Blog*, 21 December 2017, https://blogs.lse.ac.uk/impactofsocialsciences/2017/12/21/writing-a-phd-in-your-second-language-seven-reasons-youre-doing-great-and-five-ways-to-do-even-better/ [accessed 8 December 2023].
10 College of Policing, 'Delivering effective analysis', 23 October 2013, www.college.police.uk/app/intelligence-management/analysis/delivering-effective-analysis [accessed 23 February 2024].
11 Plain English Campaign, *How to Write in Plain English* (Plain English Campaign, undated), www.plainenglish.co.uk/files/howto.pdf [accessed 23 February 2024].
12 College of Policing, 'Delivering effective analysis'.
13 Andy Kirk, *Data Visualisation: A Handbook for Data Driven Design* (SAGE, 2016), p. 47.
14 *Ibid.*, p. 49.
15 Paraskevi Markou, Banu Ahtam, and Marietta Papadatou-Pastou, 'Elevated levels of atypical handedness in autism: Meta-analyses', *Neuropsychology Review*, 27 (2017), pp. 258–283, https://doi.org/10.1007/s11065-017-9354-4
16 Gavin Maxwell, *On Poetry* (Bloomsbury Academic, 2012), p. 19.

Conclusion: drawing the threads together

In this book, each chapter addresses a different format or set of formats. That may give the impression that these formats are discrete and there is no overlap between them. This is far from the case. You may have picked up that stories and storytelling underpin all the other formats, so that is one overlap. Another is that poetry comics (sometimes known as comics poetry, or graphic poetry) exist. These are not illustrated poems, or comics with dialogue that rhymes. They are a different 'hybrid form that cannot be separated out into component parts' and which are 'designed to turn your traditional linear narrative on its head'.[1] Poetry comics remove the 'sequential' from 'sequential art' which separates them from storytelling. They are often presented as a single image with no clear direction to the reader about where they might start and finish looking, or what route they might take between those two points. In this respect they are more like cartoons than comics, though there is also a debate about whether cartoons are actually a form of comics.[2] Some people say they are not because, like poetry comics, they don't have a sequential aspect. Others say they are because they combine images and text.

There are many other hybrids too. Autobiographical plays combine life writing and play writing. University textbooks can be written like novels. *Evocative Autoethnography: Writing Lives and Telling Stories*, by Arthur ('Art') Bochner and Carolyn Ellis, is written as a fictional account of a workshop run by the two of them, with characters, description, and dialogue. Here is an excerpt:

> 'I'm psyched about the energy in this room,' Art says to Carolyn, as participants engage loudly in conversation around the conference table. He pauses while they take their seats, then begins.

'We're sorry we didn't get to everyone who came up to talk with us during the break. But we can take up your comments and questions now. Barbara, I appreciated the points you raised with me during the break. Would you start us off?'

Sitting in the first row, Barbara turns to face the other participants. 'I was just saying to Art how validated I feel here. It's been my dream to do work that touches people, makes a difference, and feels meaningful to me. The whole idea of thinking like a storyteller is new to me and I know I have a lot to learn. I have to admit I've never thought of myself as a writer ...'[3]

Autoethnographers do a lot of writing, so it is perhaps not so surprising that two of them would write a creative textbook. Statisticians, though, seem like more of a stretch. Even so, *An Adventure in Statistics*, by Andy Field, is a statistics textbook, written like a speculative fiction novel, with graphic illustrations by James Iles. The main character is Zach and the book is written in the first person. Zach has a 'reality checker', a device containing an AI head called The Head which has a Californian accent. Here's an excerpt:

'If you can't help with the science, what *can* you help with?' I asked The Head.

'The maths. I was the best man at Maths' wedding to Mrs Maths, I'm godfather to baby Maths, I look after goldfish Maths when the Maths family are on holiday. Maths is my sibs. What about you?'

'I walked past Maths on a crowded street once.'

'That's real bad. Let's start real simple. What is the result of $1 + 2 + 3^2$?'

'That's easy, it's 27.'

The Head shook. 'No, it's 19.'

'Don't be crazy, 1 plus 2 is 3, multiply that by 3^2 which is 9, and you get 3 times 9, which is 27.'

'You're breaking Mr Maths' heart,' The Head said in an overly sad voice. 'You're forgetting BODMAS.'

'BODMAS? Who was he?'

'Father BODMAS was an old guy with a white beard who used to give children equations for Christmas.'[4]

In fact, as The Head goes on to explain, BODMAS is a mnemonic for the sequence of processes in an equation: brackets, order term, division, multiplication, addition, and subtraction.

I studied social psychology as an undergraduate and every Monday morning for three years we had a two-hour statistics class. To me, as a young woman, this felt like cruel and unusual punishment. The textbooks I used were dry and procedural. As the excerpt above shows, the story Andy Field tells is humorous. It is also beautifully illustrated and quite compelling to read. If I had had this book I think I would have learned to use statistics more quickly and fully. I enjoy reading stories, and Field weaves the statistical information into the story so cleverly that I would have understood it better, and remembered it more easily, because he engages my emotions and memories as well as my intelligence.

These two books are excellent examples of a point I have made several times in this book: all good writing is rooted in stories. An excellent way to learn more about stories and storytelling is to experience stories in as many forms as possible. Read poems, short stories, novels, and life stories; listen to storytellers in person and through social and mainstream media; have a go at oral storytelling; see or listen to plays, or join an amateur dramatic society or improv club; watch videos, films, and TV drama series; make short videos yourself; and so on. This is an enjoyable way to help improve your writing.

Even for those, like me, who love writing, it can't be fun all the time. Different people find different parts of the process more fun than others. Some love the discovery of writing the first draft, and find the editing work of the second draft to be eye-wateringly tedious. Others loathe churning out their raw material and take joy in crafting their work into shape. But overall, writing should be more fun than not fun. Bringing creativity into your writing process is a good way to make it more fun. Writing is always there as a resource for you, to help with your work, thoughts, or feelings, or simply for pleasure or interest. Anyone who likes puns or jokes, cares about a song lyric or the screenplay of a movie, or enjoys reading or listening to stories, can find great delight in writing creatively for their own purposes.

Notes

1 Eileen Gonzalez, 'What are poetry comics?', *Book Riot*, 23 October 2023, https://bookriot.com/poetry-comics/ [accessed 18 January 2024].
2 Nicola Streeten, 'Introduction', in *The Inking Woman: 250 Years of Women Cartoon and Comic Artists*, ed. by Nicola Streeten and Cath Tate (Myriad Editions, 2018), p. 9.
3 Arthur Bochner and Carolyn Ellis, *Evocative Autoethnography: Writing Lives and Telling Stories* (Routledge, 2016), pp. 94–95.
4 Andy Field, *An Adventure in Statistics: The Reality Enigma* (SAGE Publications, 2016), p. 49.

On being autistic

I am:
- Too soft
- Too loud
- Too shy
- Too proud
- Too sad
- Too glad
- Too much
- Not enough
- Antisocial
- An opal
- Inscrutable
- Pollutable
- Too quiet
- A pariah
- So funny
- A dummy

Index

Page numbers in *italics* denote illustrations.

abortion rights 67
acrostic poetry 65, 68
additional languages, writing in 175–177
Adventure in Statistics, An (Field) 184–185
advertising 70, 88, 90
Alocci, Tiziana 98
Alvarado, Gabriela 67
animation 20
Apple 6–7
architecture, zines in 98
Art of Statistics, The (Spiegelhalter) 49–50
arts workplaces, dramatic writing 114–115
Ash, Christopher 98
attention deficit hyperactivity disorder (ADHD) 113
audiences/readers 3, 174–175
authenticity 9, 21
autobiography 48
autoethnography 51–53, 183–184

Bannister, Roland 44–45
Barthes, Roland 47
Bayley, Stephen 34–35
Behind Closed Doors (Stark) 5, 9, 33–34
Bell, Elliot 137–138
Belleville Rendezvous (2003) 20

Bilgen-Reinart, Üstün 46
Blackmail (Snyder) 119–120
blackout poems 71–73
blogs 49, 140, 159
Bochner, Arthur 183–184
Boje, David 35
Boy with the Topknot, The (Sanghera) 50, 51–52, 55
Bussidor, Ila 46

calligraphy 85
Carroll, Lewis 66–67
Carter, Robert 64
cartoons 86–87
case files 158
case studies 54–55
cathartic writing 155–156
cause and effect 23
Chandler, Raymond 171
Cheeseman, Peter 114
Chomet, Sylvain 20
clichés 31, 125
close-ups 5, 53
coaching, comics in 90
Cohen, David 76
Cohen, Jeffrey H. 52–53
comedy writing 113, 125–126
comics 20, 88–96, *101*
  see also poetry comics
conflict 31–32

Index

constraints and creativity 75–78, 112
see also risks of creativity
construction industry, vocabularies 176
contrast 32
Conversation with a Purpose (Kara and Jackson) 92–96
correspondence, stories constructed from 24
couplets 68
creative metaphors 31
critical inner voices 167–168, 171
see also feedback
customer support 3, 115
cut-up poems 71

Davidson, Jonathan 46–47, 138–139
dead metaphors 31
dialogue 25–29, 112
diaries
 private 3
 workplace 157–159
 see also journals
digital comics, making 91–92
digital writing 134–143
digital zines 96
distractions 171–172
documentary theatre 114–115
dramatic writing 112–130
 comedy 113, 125–126
 play scripts and screenplays 113–125
Drnaso, Nick 88
Duval, Marie 86, 87

education
 comics 88, 89, 91–96
 dramatic writing 113–114
 vocabularies 176
Eisner, Will 88
Elbow, Peter 132–133, 151–153
Ellis, Carolyn 51, 183–184
emails 134–139

emotions 47–48, 149, 155, 156, 174
see also sensory language
epistolary structures 24
epistolary writing *see* digital writing; letters
ethical life writing 57
Ethiopia 61
Evocative Autoethnography (Bochner and Ellis) 183–184
Exercises in Style (Queneau) 35
ezines 96

fanzines 96
feedback 172–174
fiction *see* stories/fiction
Field, Andy 184–185
film shots 52
found poems 71–75
Fox, Kate 126, 138
free verse 70–71, 75
freewriting 150–151
 directed 152–154
friend, writing's role as 156–157

Goldstein, Tara 114
Google Chrome 89
grammar 165
Graphic Medicine International Collective 89
graphic writing 85–111
 cartoons 86–87
 comics 20, 88–96, *101*
 zines 96–100, *102–106*
Great Escape, The (Soni) 5–6, 7–8, 24–25, 28–29

haiku 68–69
handover notes 158
Hanff, Helene 24
hard copy zines *97*, 98
Harron, Piper 9–10
healthcare 3, 89, 115, 140
Herriot, James 48
Holloway, Sally 126
Holub, Miroslav 61

home–school partnerships, dramatic writing 113
Hot Fuzz (Wright and Pegg) 120–122

impostor syndrome 171
improvisation 112–113
instructional comics 88, 89, 91
    see also education
intelligence analysis (UK College of Policing) 177
internet 140–142
I-poems 73–74
iTunes 6–7

Jackson, Sophie 96
Japanese haiku 68–69
jargon 176–177
'Jewish Food' (Larkin) 77–78
John, Zoë 68
jokes 125, 126
journalism
    and comics 88–89
    and social media 140
journals
    reflective 154–155
    workplace 157–159

keyword poems 74–75
Kim, Henn 73–74
Kovalevskaya, Sofia Vasilyevna 61

Lacks, Henrietta 24
Larkin, Joan 77–78
letters 131–134, *135–137*, 144–145
life writing 44–60
    autoethnography 51–53
    learning to read 57–58
    memoir 48–51
    observational writing 45–48
    reports and case studies 53–55
    rights and wrongs in 56–57
    role of theory in 55–56
Life's a Pitch (Bayley and Mavity) 34–35

limericks 9, 69, 76
lists 149
long forms of writing 151

making comics 91–96
making zines 99
Manners, Jordan 114
Māori tattoos 85
Market Cafe Magazine 97, 98
marketing 70, 88, 90
Mars (confectionery company) 90
Mavity, Roger 34–35
Maxwell, Glyn 64, 152
McCloud, Scott 89
McKee, Robert 118–119
medicine *see* healthcare
memoir 48–51, 89
metaphors 30–31
'Metro' (Davidson) 46–47
mnemonics 65
monologue 112
Morgan, Nicola 27, 28
myths about writing 166–168

non-fiction
    creative writing in 4–7
    suspense 32–33
novels, graphic 88
Nunumi 20
Nyabola, Nanjala 22

observational writing 45–48
Okrent, Arika 2–3
online writing *see* digital writing
Osaka, Naomi 75
out-of-office messages 138–139

Parker, Robert B. 26–28
Pegg, Simon 120–122
Pennebaker, James 155
perzines 96
plain language 176–177
play scripts 113–126
PO Box zines 98
'Poems on the Underground' 61

Index

poetry 46–47, 61–84
  constraints and creativity 75–78
  found poems 71–75
  out-of-office messages 138–139
  poetic forms 65–70, 71
  what is a poem? 63–65
  workplaces, poetry in 61, 70–71
poetry comics 183
policing 177
post-traumatic stress disorder 154
Powell, Padgett 24
Powers, Gary 25–26
Pratchett, Terry 7
productivity 168–172
Project Orange 98
pronouns, in emails 139
public information comics 90

quatrains 69–70
Queneau, Raymond 35

Rajan, Amol 139
readers/audiences 3, 174–175
reading out loud 166
recapitulation 30, 34
reflective journaling 154–155
reflexivity 57
repetition 29–30, 34
reports 53–54
risks of creativity 143
rituals 172
role play 112–113
Rudd, Dale 25–26

sales 115
Sanghera, Sathnam 50, 51–52, 55
Sayisi Dene people 45–46
scene writing, play scripts and screenplays 118–122, 123–124
schools *see* education
science/STEM careers 134
scientist-poets 61
Scottish Centre for Comics Studies 90
screenplays 115–129
Segura (design and communications firm) 10–11
self-publishing and zines 96
sensory language 6, 7, 33–34
sentence structure, variation in 29, 32
sequential art 85–86
  see also comics
sestinas 70, 76–78
shaped poetry 66–68
Shea, Virginia 142
Siirola, Jeffrey 25–26
Sikoryak, Robert 6–7
Skloot, Rebecca 24
Sky Rover (2016) 20
Slater, Nigel 50–51
Snyder, Lynn 119–120
So Fi Zine 98
Soane, James 98
social media 139–140
social science zines *see So Fi Zine*
social work 158
Soni, Saket 5–6, 7–8, 24–25, 28–29
sonnets 70
specialist vocabularies 176–177
speech, grammar of 165
speech tags 28
Spiegelhalter, David 49–50
Spiegelman, Art 88
Stark, Laura 5, 9, 33–34
statistics 49–50, 184–185
Stein, Sol 27
STEM careers 134
stories/fiction 20–43
  structure 23–25
  writing devices 25–35
structure 23–25, 29
suspense 32–33
synonyms 29–30

teacher, writing's role as 150–154
tech companies, comics in 89
tension, creating 7–8
'Text in the City' 61, 62
theatrical techniques *see* dramatic writing

theory, role of, in life writing
55–56
therapist, writing's role as 154–156
Toast (Slater) 50–51
training *see* education;
instructional comics
transition, managing a 5
translation 175–176
trauma 154, 155
truth 8–9
two kinds of 21
Twain, Mark 44

Ukraine 46–47
urban poetry 61, 62

Van Every, Jo 139
verbatim theatre 114–115
Victoria Theatre, Stoke-on-Trent
114–115
villanelle 70
visualisation 167
vocabularies 176–177
voice(s) 32, 35

Watson, Ash 98
Watson, Cate 113
websites 141–142

Wertenbaker, Timberlake
118, 119
Wickersham, Joan 24
Wickett, Jocelyn 114
Williams, Ian 89
Wood, James 21
workplaces
audiences 174–175
cartoons in 86
comics in 88–91
conflict in 32
diaries in 157–159
feedback 172–174
letter writing and digital writing
in 132–143
memoirs of 48–49
play scripts and screenplays in
115–125
poetry in 61, 70–71
productivity 168–172
role play in 112–113
see also arts workplaces;
education; healthcare
Wright, Edgar 120–122
writing devices 25–35

Zagami, Piero 98
zines 96–100, *102–106*

EU authorised representative for GPSR:
Easy Access System Europe, Mustamäe tee 50,
10621 Tallinn, Estonia
gpsr.requests@easproject.com

www.ingramcontent.com/pod-product-compliance
Lightning Source LLC
Chambersburg PA
CBHW070356240426
43671CB00013BA/2526